D0845183

# Warming the World

# Tables

# Contents

© 2000 Massachusetts Institute of Technology

This book was set in Palatino by Best-set Typesetter Ltd., Hong Kong
Printed and bound in the United States of America.

Library of Congress Cataloging-in-Publication Data

Nordhaus, William D.
  Warming the world : economic models of global warming / William D. Nordhaus and
Joseph Boyer.
    p.   cm.
  Includes bibliographical references and index.
  ISBN 0-262-14071-3 (hc. : alk. paper)
    1. Global warming—Economic aspects—Mathematical models. 2. Economic
development—Environmental aspects—Mathematical models. I. Boyer, Joseph, 1969–
II. Title.

QC981.8.G56.N67   2000
363.738'74—dc21                                                                00-029230

# Warming the World

Economic Models of
Global Warming

William D. Nordhaus and
Joseph Boyer

The MIT Press
Cambridge, Massachusetts
London, England

# Figures

# Preface

Dealing with complex scientific and economic issues has increasingly involved developing scientific and economic models that help analysts and decision makers understand likely future outcomes as well as the implications of alternative policies. This book presents the details of a pair of integrated-assessment models of the economics of climate change. The models, called RICE-99 (for the Regional Dynamic Integrated model of Climate and the Economy) and DICE-99 (for the Dynamic Integrated model of Climate and the Economy), build upon earlier work by Nordhaus and collaborators, particularly the DICE and RICE models constructed in the early 1990s. The purpose of this book is to lay out the logic and details of RICE-99 and DICE-99. Like an anatomy class, this description highlights internal structure of the models and the ways different segments are connected.

The book is organized into two parts. The first part describes RICE-99 and its globally aggregated companion, DICE-99. This part contains an introduction (chapter 1) and a brief description of RICE-99 (chapter 2) that includes all the model equations. The details of the derivation of these equations and their parameterization are presented in chapters 3 and 4. Chapters 1 through 4 present RICE-99, leaving explicit discussion of DICE-99 to chapter 5. Chapter 6 explains how the models are solved. Part II presents the major results of RICE-99 and applies it to the questions surrounding climate change. The appendixes provide a summary listing of the equations, a variable list, and the programs for the RICE-99 and DICE-99 models. The models and spreadsheets are also available on the Web.

Those interested in this exciting field will recognize that this book builds on earlier work of the authors and of many others. Although it bears the names of two authors, the intellectual inspiration and contribution of many should be recognized. Among those we thank for contributing directly or indirectly are Jesse Ausubel, Howard Gruen-specht, Henry Jacoby, Dale Jorgenson, Charles Kolstad, Alan Manne, Robert Mendelsohn, Nebojsa Nakicenovic, John Reilly, Richard Richels, Thomas Schelling, Richard Schmalensee, Stephen Schneider, Leo Schrattenholzer, Robert Stavins, Ferenc Toth, Karl Turekian, Paul Waggoner, John Weyant, Zili Yang, and Gary Yohe. Megan McCarthy and Ben Gillen provided valuable research assistance. This research was supported by the National Science Foundation and the Department of Energy. None of these is responsible for the errors, opinions, or flights of fancy in this work.

# I

Developing the RICE and DICE Models

# 1    Introduction

"God does not play dice with the universe," was Albert Einstein's reaction to quantum mechanics. Yet humanity *is* playing dice with the natural environment through a multitude of interventions: emitting into the atmosphere trace gases like carbon dioxide that promise to change the global climate, adding ozone-depleting chemicals, engineering massive land-use changes, and depleting multitudes of species in their natural habitats, even as we create in the laboratory new organisms with unknown properties. In an earlier era, human societies learned to manage—or sometimes failed to learn and mismanaged—the grazing or water resources of their local environments. Today, as human activity increasingly affects global processes, we must learn to use wisely and protect economically our common geophysical and biological resources. This task of understanding and controlling interventions on a global scale is *managing the global commons*.

Climatologists and other scientists warn that the accumulations of carbon dioxide ($CO_2$) and other greenhouse gases (GHGs) are likely to lead to global warming and other significant climatic changes over the next century. This prospect has been sufficiently alarming that governments have undertaken, under the Kyoto Protocol of December 1997, to reduce their GHG emissions over the coming years. The Kyoto Protocol raises a number of fundamental issues: Are the emissions limitations proposed there sufficient, insufficient, or excessive? Is the mechanism proposed to combat global warming—limiting emissions from high-income countries—workable and desirable? Was it wise to omit developing countries? Is there a trajectory for the Kyoto Protocol that will lead to a comprehensive climate-change policy? Are other approaches, such as harmonized carbon taxes or geoengineering, worth considering? How does the approach in the Kyoto Protocol compare with the economist's dream of an "efficient" policy? And, perhaps most important, will these costly approaches sell in the political marketplace of the world's democracies and oligarchies?

Natural scientists have pondered many of the *scientific* questions associated with greenhouse warming for a century. But the *economic, political, and institutional* issues have only begun to be considered over the last decade. The intellectual challenge here is daunting—raising formidable issues of data, modeling, uncertainty, international coordination, and institutional design. In addition, the economic stakes are enormous. Several recent studies of the Kyoto Protocol put the price tag on abatement to be around $1 trillion in present value.[1] It is no hyperbole to say that the issue of greenhouse warming invokes the highest form of global citizenship—where nations are being called upon to sacrifice hundreds of billions of dollars of present consumption in an effort that will largely benefit people in other countries, where the benefit will not come until well into the next century and beyond, and where the threat is highly uncertain and based on modeling rather than direct observation.

The issue of global warming has proven one of the most controversial and difficult problems facing nations as they cross the bridge into the twenty-first century. Over the last decade, the issue has migrated from the scientific journals to White House conferences and world summit meetings. In response, a small navy of natural and social scientists has been mobilized to help improve our understanding. In parallel with the growing interest, industrial, environmental, and political groups have put their oars in the water to pull the ship in directions favorable to their ideologies or bottom lines.

Among the most impressive advances over the last decade has been the development of integrated-assessment economic models that analyze the problem of global warming from an economic point of view. Literally dozens of modeling groups around the world have brought to bear the tools of economics, mathematical modeling, decision theory, and related disciplines. Whereas a decade ago, not a single integrated dynamic model of the economics of climate change existed, there are now more than we can keep track of.

One of the earliest dynamic economic models of climate change was the DICE model (a Dynamic Integrated model of Climate and the Economy). Originally developed from a line of energy models, DICE integrated in an end-to-end fashion the economics, carbon cycle,

---

1. See the studies contained in Weyant 1999.

**Table 1.1**
Reference case output across model generations for the year 2100

|  | DICE-94 | RICE-99 |
|---|---|---|
| Industrial emissions (GtC/year)[1] | 24.9 | 12.9 |
| Output (trillions of 1990 U.S.$) | 111.5 | 97.02 |
| Population (billions) | 9.8 | 10.7 |
| Output per person (thousands of 1990 U.S.$) | 11.4 | 9.1 |
| Carbon intensity (tons carbon per $1,000 of GDP, 1990 U.S.$) | 0.22 | 0.13 |
| Temperature (degrees C above 1900) | 3.39 | 2.42 |

Note: 1. "GtC" means billion metric tons of carbon.
Source: Runs of models as described in text.

climate science, and impacts in a highly aggregated model that allowed a weighing of the costs and benefits of taking steps to slow greenhouse warming. The first version of DICE was presented in 1990, and the results of the full model were described in Nordhaus 1994b. A regionalized version, known as RICE (a Regional dynamic Integrated model of Climate and the Economy), was developed and presented in Nordhaus and Yang 1996.

Although the basic structure of the DICE and RICE models has survived in the crucible of scientific criticism, further developments in both economics and the natural sciences suggest that major revisions of the earlier approaches would be useful. Although no simple solutions have been found, a number of small discoveries and large innovations in the natural and social sciences have come forth. Moreover, the past decade has seen major improvements in the underlying data on greenhouse-gas emissions and energy and economic data.

This book represents the fruits of the revision of the earlier models. The new models have benefited from a thorough overhaul while maintaining their basic structure. Table 1.1 compares projections of the major variables in RICE-99 with the earlier DICE-94 model for the reference case in 2100.[2]

The major changes from the old to the new models are the following:

1. The major methodological change is a respecification of the production relations. Whereas the earlier DICE and RICE models used a

2. The reference case represents the model's projections for what will happen if no government control over GHG emissions is imposed. See chapter 2, section four, or chapter 6 for more complete definition.

parameterized emissions–cost relationship, the new RICE model use a three-factor production function in capital, labor, and carbon-energy. The new RICE model develops an innovative technique for representing the demand for carbon fuels and uses existing energy-demand studies for calibration.

2. The new RICE model changes the treatment of energy supply to incorporate the exhaustion of fossil fuels. This approach treats the supply of fossil fuels explicitly and uses a market-determined process to drive the depletion of exhaustible carbon fuels. The new model incorporates a depletable supply of carbon fuels, with the marginal cost of extraction rising steeply after 6 trillion tons of carbon emissions.[3] (This would be the equivalent of burning about 9 trillion tons of coal.) With limited supplies, fossil fuel prices will eventually rise in the marketplace to choke off consumption of fossil fuels.

3. Most of the data have been updated by almost a decade to reflect data for 1994–98. The output growth in the models is generated by regional economic, energy, and population data and forecasts. The new model projects significantly lower reference $CO_2$ emissions over the next century than the earlier DICE and RICE models because of slower projected growth and a higher rate of decarbonization of the world economy.

4. The RICE/DICE-99 carbon-cycle model is now a three-box model, with carbon flows among the atmosphere, upper biosphere-shallow oceans, and deep oceans. (In earlier versions, carbon simply disappeared at a constant rate from the atmosphere.) The temperature dynamics in the new models remain unchanged from the earliest versions, as climate research has not produced compelling reasons to alter them. Forcings from non-$CO_2$ GHGs, and aerosols have been updated to reflect more recent projections. The projected global temperature change in the reference case turns out to be significantly lower in the current version of RICE. This is due to the inclusion of negative forcings from sulfates in RICE-99, the lower forcings from the chlorofluorocarbons, and the slower growth in $CO_2$ concentrations (see table 1.2).

---

3. We sometimes refer to carbon dioxide emissions and concentrations as "carbon emissions," "concentrations of carbon," or sometimes simply "emissions" or "concentrations." Both are measured in metric tons of carbon. We refer to metric tons of carbon as simply "tons of carbon." In some contexts, as noted, particularly when referring to coal, "tons" will mean short tons.

**Table 1.2**
Difference in radiative forcing across models, reference case, 2100

|  | Watts per m$^2$ | Percent of total difference |
|---|---|---|
| *Total difference (RICE-99 minus DICE-94)* | −1.73 | 100.00 |
| *Carbon emissions and cycle* |  |  |
| Carbon emissions (GtC/year)[1] | −0.89 | 51.26 |
| Starting carbon concentration | −0.06 | 3.25 |
| Carbon cycle model | 0.30 | −17.49 |
| *Other anthropogenic forcings* |  |  |
| Chlorofluorocarbons | −0.42 | 24.28 |
| Sulfate aerosols | −1.06 | 61.27 |
| Other greenhouse gases | 0.45 | −26.01 |
| *Change in preindustrial carbon concentration in forcing equation* | −0.06 | 3.43 |

Note: 1. "GtC" means billion metric tons of carbon.

5. The impacts of climate change have been revised significantly in the new models. The global impact is derived from regional impact estimates. These estimates are derived from an analysis that considers market, nonmarket, and potential catastrophic impacts. The resulting temperature damage function is more pessimistic than that of the original DICE model.

6. The RICE and DICE models were originally developed using the General Algebraic Modeling System software package. The new versions have been programmed both in GAMS and in an EXCEL spreadsheet version so that other researchers can easily understand and use the models.

This book lays out the revisions and their implications in detail. The underlying philosophy of the original DICE and RICE models remains unchanged: to develop small and transparent models that can be easily understood, can be modified as new data or results emerge, and will be useful for scientific, teaching, and policy purposes.

It is our hope that this book can help modelers and policymakers better understand the complex trade-offs involved in climate-change policy. In the end, good analysis cannot dictate policy, but it can help policymakers thread the needle between a ruinously expensive climate-change policy that today's citizens will find intolerable and a do-nothing policy that the future will curse us for.

# 2    The Structure and Derivation of RICE-99

## Overview of Approach

This chapter presents an overview of RICE-99. The first section describes the structure of the model, while subsequent sections present the equations of the model. The following chapter then discusses the calibration of the major components of the models.

In considering climate-change policies, the fundamental trade-off that society faces is between consumption today and consumption in the future. By taking steps to slow emissions of greenhouse gases today, the economy reduces the amount of output that can be devoted to consumption and productive investment. The return for this "climate investment" is lower damages and therefore higher consumption in the future. The climate investments involve reducing fossil-fuel consumption or moving to low-carbon fuels; in return for this investment, the impacts on agriculture, coastlines, and ecosystems as well as the potential for catastrophic climate change will be reduced.

But the lags between emissions reductions and climatic impacts are extraordinarily long and uncertain which makes the economic and scientific questions treacherous. Nations must decide whether they will take climate investments now to slow climate change over the coming centuries. Few societal decisions and no personal ones, except those involving Pascal's wager, have comparable time horizons, and this encourages political decision makers to temporize on costly steps.

The major challenge in RICE-99 has been to develop a model of the world economy that captures the significant properties of medium- and long-run economic growth of the major countries and regions over the next century. Outside of the rarified and highly stylized models used in the climate-change integrated-assessment models, there are essentially no models of the world economy upon which to draw. Useful ingredients can be obtained for the population projections from demographers, who do in fact prepare long-term projections. But for other important variables, ones determining capital formation and technological change, particularly for countries outside the United States and Western Europe, it has been necessary to develop long-term projections *de novo*.

The model operates in periods of ten years. All flow variables in the empirical model are reported as flows per year, while the convention is that stocks are measured at the beginning of the period.

## Model Description

### Economic Sectors

The approach taken here is to view climate change in the framework of economic growth theory. This approach was developed by Frank Ramsey in the 1920s (see Ramsey 1928), made rigorous by Tjalling Koopmans and others in the 1960s (see especially Koopmans 1967), and is summarized by Robert Solow in his masterful exposition of economic-growth theory 1970. In the neoclassical growth model, society invests in tangible capital goods, thereby abstaining from consumption today to increase consumption in the future.

The DICE-RICE models are an extension of the Ramsey model to include climate investments in the environment. Emissions reductions in the extended model are analogous to investment in the mainstream model. That is, we can view concentrations of GHGs as "negative capital," and emissions reductions as lowering the quantity of negative capital. Sacrifices of consumption that lower emissions prevent economically harmful climate change and thereby increase consumption possibilities in the future.

The description that follows focuses on the fully regionalized model, the RICE model. Most of the statements apply equally to the globally aggregated DICE model, which is discussed in chapter 5.

The world is composed of several regions. Some regions consist of a single sovereign country (such as the United States or China) while other regions (like OECD Europe or the low-income region) contain many countries. Each region is assumed to have a well-defined set of preferences, represented by a "social welfare function," which determines choices about the path for consumption and investment. The social welfare function is increasing in the per capita consumption of each generation, with diminishing marginal utility of consumption. The importance of a generation's per capita consumption depends on its relative size. The relative importance of different generations is affected by a pure rate of time preference; because a positive time preference is assumed, current generations are favored over future generations.

Regions are assumed to maximize the social welfare function subject to a number of economic and geophysical constraints. The decision variables that are available to the economy are consumption, the rate of investment in tangible capital, and the climate investments, primarily reductions of GHG emissions.

The model contains both a traditional economic sector found in many economic models and a novel climate sector designed for climate-change modeling. The traditional sector of the economy—the economy without any considerations of climate change—is first described.

Each region is assumed to produce a single commodity that can be used for either consumption or investment. In the model, all changes in welfare, including those due to climate change, are included in the definition of consumption of this single commodity. Thus, we will sometimes refer to consumption of this all-inclusive commodity as "generalized consumption."

There is no international trade in goods or capital except in exchange for carbon emissions permits. Thus regions are allowed to trade only for the sake of paying other regions to lower their emissions or to receive payment for lowering emissions.

Each region is endowed with an initial stock of capital and labor and an initial and region-specific level of technology. Population growth and technological change are exogenous while capital accumulation is determined by optimizing the flow of consumption over time.

The major methodological change in the economic sector is a respecification of the production relations in RICE from earlier vintages. (DICE retains the simple reduced-form production structure from earlier vintages.) RICE-99 defines a new input into production called *carbon-energy*. Carbon-energy can be thought of as the energy services derived from fossil-fuel consumption. Fossil-fuel consumption in the model is equal to the carbon content of all fossil-fuel consumption. In other words, energy use is lumped into a single aggregate where the different fuels are aggregated using carbon weights. Thus we refer to the marginal product or cost of, supply of, and allocation across regions of carbon-energy rather than coal, petroleum, and natural gas.

Output is produced with a Cobb-Douglas production function in capital, labor, and carbon-energy inputs. Technological change takes two forms: economy-wide technological change and carbon-saving technological change. Economy-wide technological change is Hicks-neutral, while carbon-saving technological change is modeled as reducing the ratio of $CO_2$ emissions to carbon-energy inputs. For convenience, both carbon-energy and industrial emissions are measured in carbon units. The procedure is quite intuitive if one thinks of carbon-energy as coal.

The energy-related parameters are calibrated using data on energy use, energy prices, and energy-use price elasticities. These allow a empirically based carbon reduction curve, whereas most current integrated assessment models make reasonable but not data-based specifications of demand. On the supply side, the earlier DICE and RICE models assumed that carbon fuels are superabundant at a fixed supply price. In RICE-99, a supply curve for carbon-energy is introduced. The supply curve allows for limited (albeit huge) long-run supplies at rising costs. Because of the optimal-growth framework, carbon-energy is efficiently allocated across time, which implies that low-cost carbon resources have scarcity prices (called *Hotelling rents*) and that carbon-energy prices rise over time.

### Climate-Related Sectors

The nontraditional part of the model contains a number of geophysical relationships that link the different forces affecting climate change. This part contains a carbon cycle, a radiative forcing equation, climate-change equations, and a climate-damage relationship.

In the earlier DICE-RICE models, endogenous emissions included $CO_2$ and CFCs. In RICE-99 and DICE-99, endogenous emissions are limited to industrial $CO_2$. This reflects projections by the Intergovernmental Panel on Climate Change (IPCC) and within the DICE/RICE models that indicate the radiative forcings from uncontrolled $CO_2$ concentrations are likely to be nearly five times larger than those from the combined effect of non-$CO_2$ GHGs and aerosols (see table 3.9, which is discussed in chapter 3). The major change here is that the chlorofluorocarbons (CFCs) are now outside the climate-change control strategy; this specification reflects the fact that CFCs are strictly controlled outside the framework of the climate-change agreements under different protocols. Other anthropogenic contributions to climate change are also taken as exogenous. These include $CO_2$ emissions from land-use changes, non-$CO_2$ GHGs, and sulfate aerosols.[1] Although it would be more complete to endogenize other GHGs and aerosols (and five other gases are in principle included in the Kyoto Protocol), these are extremely complex and poorly understood.

The original DICE and RICE models used an empirical approach to estimating the carbon flows, estimating the parameters of the emissions-concentrations equation from data on emissions and concentrations. A number of commentators noted that this approach may understate the long-run atmospheric retention of carbon because it assumes an infinite sink of carbon in the deep oceans. DICE-99 and RICE-99 replace the earlier treatment with a structural approach that uses a three-reservoir model calibrated to existing carbon-cycle models. The basic idea is that the deep oceans provide a finite sink for carbon in the long run. In the new specification, we assume that there are three reservoirs for carbon—the atmosphere, a quickly mixing reservoir in the upper oceans and the short-term biosphere, and the deep oceans. Carbon flows in both directions between adjacent reservoirs. The mixing between the biosphere/shallow ocean reservoir and the deep oceans is extremely slow. The RICE-DICE-99 approach matches the original DICE model and other calculations in the early periods but has better long-run properties. A full discussion of this new approach is contained in chapter 4.

Climate change is represented by global mean-surface temperature, and the relationship uses the consensus of climate modelers and a lag suggested by coupled ocean-atmospheric models. The climate module is unchanged from the original DICE and RICE models.

Understanding the economic impacts of climate change continues to be the thorniest issue in climate-change economics. Estimates of climate-change impacts in most integrated assessment modeling rely on a wide variety of estimates of the damage from climate change in different sectors for different regions. Starting with Nordhaus 1989, 1991a, assessments tended to organize impacts of climate change in the framework of national economic accounts, with additions to reflect nonmarket activity. This book follows first-generation approaches by analyzing impacts on a sectoral basis. There are two major differences

---

1. Although total carbon emissions include both industrial and land emissions, often we will refer to the endogenous component, industrial emissions, as simply "emissions" or "carbon emissions."

here from many earlier studies. First, the approach focuses on deriving estimates for all regions rather than for the United States alone. This focus is necessary both because global warming is a global problem and because the impacts are likely to be significantly larger in poorer countries. Second, this book focuses more heavily on the nonmarket aspects of climate change with particular importance given to the potential for catastrophic risk. This approach is taken because of the finding of the first-generation studies that the impacts on market sectors are likely to be relatively limited. The major results are that impacts are likely to differ sharply by region. Russia and other high-income countries (principally Canada) are likely to benefit slightly from a modest global warming. At the other extreme, low-income regions—particularly Africa and India—and Western Europe appear to be quite vulnerable to climate change. The United States appears to be relatively less vulnerable to climate change than many countries. These results are discussed in detail in chapter 4.

### Derivation of the Equations of RICE-99

The equations of RICE-99 are discussed here in detail. The relationships are divided into three groups: the objective function, the economic relationships, and the geophysical relationships. Although the economic sectors are conventional in their approach, modifying them for the climate-change problem requires careful attention, and the major issues are considered in the first two subsections. The major issues of the climate sector and the interaction of economy and climate are analyzed in third subsection.

### *Objective Function*

A central organizing framework of the DICE-RICE models is that the purpose of economic and environmental policies is to improve the living standards or consumption of people now and in the future. The relevant economic variable is *generalized consumption*, which denotes a broad concept that includes not only traditional market purchases of goods and services like food and shelter but also nonmarket items such as leisure, cultural amenities, and enjoyment of the environment.

The fundamental assumption adopted here is that policies should be designed to optimize the flow of generalized consumption over time. This approach rests on the view that more consumption is preferred to less. Moreover, increments of consumption become less valuable as consumption levels increase. In technical terms, we model these assumptions by assuming that regions maximize a social welfare function that is the discounted sum of the population-weighted utility of per capita consumption. This social welfare function is a mathematical representation of three basic value judgments: (1) higher levels of consumption have higher worth; (2) there is diminishing marginal valuation of consumption as consumption increases; and (3) the marginal social utility of consumption is higher for the current generation than for a future generation of the same size with the same per capita consumption.

RICE adds a significant level of complexity to the original DICE model by incorporating the simultaneous growth paths of different regions. The exact objective function, or criterion to be maximized, for region $J$ is:

$$W_J = \sum_t U[c_J(t), L_J(t)]R(t), \tag{2.1}$$

where $W_J$ is the objective function of region $J$, $U[c_J(t),L_J(t)]$ is the utility of consumption for region $J$, $c_J(t)$ is the flow of consumption per capita during period $t$, $L_J(t)$ is the population at time $t$, and $R(t)$ is the pure time preference discount factor. The exact form of the utility function will be described shortly.

Utility is discounted by a factor that represents social time preference among different generations. The pure rate of time preference $\rho(t)$, which underlies the time preference discount factor $R(t)$, becomes an important parameter in this approach; the parameter $\rho(t)$ is assumed to decline over time, and the pure time preference discount factor is then given by:

$$R(t) = \prod_{v=0}^{t}[1+\rho(v)]^{-10}. \tag{2.2}$$

The pure rate of time preference is a choice parameter that is implicit in many societal decisions, such as fiscal and monetary policies. In conjunction with other parameters, it is closely connected with the market rate of interest (or marginal productivity of capital) and with the savings rate. The original RICE and DICE models used a constant pure rate of time preference of $\rho(t) = 3$ percent per year. The constant rate of 3 percent per year was considered to be consistent with historical savings data and interest rates. In DICE-99 and RICE-99, the pure rate of time preference is assumed to decline over time because of the assumption of declining impatience. The rate of time preference starts at 3 percent per year in 1995 and declines to 2.3 percent per year in 2100 and 1.8 percent per year in 2200.[2]

### Economic Constraints

The next set of equations represents the different regions. The first equation is the definition of utility, which was described and motivated in the previous subsection. Utility represents the current value of economic well-being and is assumed equal to the size of population $[L_J(t)]$ times the utility of per capita consumption $u[c_J(t)]$. Equation (2.3) uses the general case of a power function to represent the form of the utility function:[3]

$$U[c_J(t), L_J(t)] = L_J(t)\{c_J(t)^{1-\alpha} - 1\}/(1 - \alpha). \tag{2.3}$$

In this equation, the parameter $\alpha$ is a measure of the social valuation of different levels of consumption, which has several interpretations. It represents the curvature of the utility function, the elasticity of the marginal utility of consumption, or the rate of inequality aversion. Operationally, it measures the extent to which a region is willing to reduce the welfare of high-consumption generations to improve the welfare of low-consumption generations. In the RICE and DICE models, we take (the limit of) $\alpha = 1$, which yields the logarithmic or Bernoullian utility function:

$$U[c_J(t), L_J(t)] = L_J(t)\{\log[c_J(t)]\}. \tag{2.3'}$$

For most regions, the growth of population is assumed to follow an exponential path, and the basic projection method is as follows: Population growth in the initial period is taken from U.N. data, as discussed below. We then assume that the growth rate declines over time at a geometrically declining rate. More precisely, let $g^{\text{pop}}_J(t)$ be the population growth rate in region $J$ and period $t$, and $\delta^{\text{pop}}_J$ be its constant rate of decline. The growth rate of population in time $t$ is then:

---

2. A comprehensive review of issues involved in discounting the distant future is contained in the essays in Portney and Weyant 1999. A full discussion of the discount rate question in the context of the DICE and RICE models is contained in Nordhaus 1994b and 1998a.

3. This formulation subtracts one from the power function in the numerator of equation (3.1) so that the limit of the expression is $L_J(t)[\log(c_J(t))]$ as $\alpha$ tends to 1.

$$g^{\text{POP}}_J(t) = g^{\text{POP}}_J(0)\exp(-\delta^{\text{POP}}_J t). \tag{2.4}$$

It is easily verified that this assumption leads to a stable population. Its advantage is that the population trajectory can be represented by two parameters and can be easily fit to different projections. The parameters chosen for RICE-99 produce a global population growth rate of 1.5 percent per year for the initial decade and a rate of decline in the global population growth rate of about 20 percent per decade. The global asymptotic maximum population is 11.5 billion people.

Production is represented by a modification of a standard neoclassical production function. For region $J$, output or GDP [$Q_J(t)$] is given by a constant-returns-to-scale Cobb-Douglas production function in capital [$K_J(t)$], labor [$L_J(t)$], and carbon-energy. $ES_J(t)$. Carbon-energy represents energy services. Carbon emissions is related to energy services by an efficiency index function; this function changes over time to reflect carbon-saving technological change.

$$Q_J(t) = \Omega_J(t)\{A_J(t)K_J(t)^{\gamma} L_J(t)^{1-\beta_J-\gamma} ES_J(t)^{\beta_J} - c^E_J(t)ES_J(t)\}. \tag{2.5a}$$

$$ES_J(t) = \varsigma_J(t)E_J(t). \tag{2.5b}$$

In equation (2.5a), $\gamma$ is the elasticity of output with respect to capital and is assumed to be 0.3. $\beta_J$ is the elasticity of output with respect to energy services (discussed below), and the term $(1 - \beta_J - \gamma)$ is the output elasticity with respect to labor. $A_J(t)$ represents the level of Hicks-neutral technological change. The term $\Omega_J(t)$ is a damage coefficient that relates to the impact of climate change on output and is described below. Labor inputs are equal to population; this is identical to assuming they are proportional to population and adjusting $A_J(t)$ by a constant factor. Capital accumulation is described below, and the carbon-energy aggregate is discussed next. The term [$c^E_J(t)ES_J(t)$] in equation (2.5a) subtracts from gross output the costs of producing carbon-energy.

Equation (2.5b) then shows the relationship between carbon-energy inputs and energy services. Technological change in the energy sector is carbon-augmenting, where $\varsigma_J(t)$ is the level of carbon-augmenting technology. Because of carbon-augmenting technological change, society is able to squeeze more energy services per unit of carbon-energy.

A major uncertainty in the model involves projecting the growth of $A_J(t)$, or total factor productivity (TFP), into the future. TFP growth is assumed to slow gradually over the next three centuries until eventually stopping. The exact technique for deriving estimates is described in chapter 3, the third section, the first subsection. The technical formula within the DICE and RICE models for projecting TFP is similar to that introduced above for population growth. Let $g^A{}_J(t)$ be the TFP growth rate in period $t$ and $\delta^A{}_J$ be its constant rate of decline. Productivity growth at time $t$ is then:

$$g^A{}_J(t) = g^A{}_J(0)\exp(-\delta^A{}_J t),\tag{2.6}$$

where $\delta^A{}_J$ is chosen so that $A_J(t)$ tends asymptotically to $A_J^*$, where $A_J^*$ is the assumed asymptotic level of total factor productivity for region $J$.

In a one-sector closed economy $Q_J(t)$ equals $C_J(t) + I_J(t)$, where $C_J(t)$ is consumption and $I_J(t)$ is investment. In RICE-99, regions can trade carbon emissions permits for goods. With trade, the constraint on regional expenditures becomes:

$$Q_J(t) + \tau_J(t)[\Pi_J(t) - E_J(t)] = C_J(t) + I_J(t),\tag{2.7}$$

where $\Pi_J(t)$ is the number of carbon emissions allowances allocated to region $J$ and $\tau_J(t)$ is the price of each emissions permit. The second term on the left-hand side of equation (2.7) measures the net revenues a region receives from its purchase and sale of permits. If its emissions exceed its allocation of permits, it has to buy more permits than it sells, and its net revenue is negative. We will refer to $\tau_J(t)$ below as the *carbon tax*, because it functions just like a tax on carbon, but it can also be interpreted as the market price of emissions permits. The allocation of emissions permits is determined by agreement among the parties. Each region also takes the carbon tax to be exogenous.

A central research and policy issue is the number and composition of emissions trading blocs. A trading bloc $B$ is a set of regions for which the carbon tax (or permit price) is equalized and for which total emissions cannot exceed the total allocation of permits. Grouping regions into trading blocs makes it easy to analyze the impacts of policies such as the Kyoto Protocol that call for emissions trading. Equation set (2.7') gives the mathematical conditions for the permit allocations and carbon taxes in a trading bloc:

$$\tau_J(t) = \tau_b(t) \text{ for all } J \text{ who are members of } B$$

$$\sum_{J \in B} \Pi_J(t) \geq \sum_{J \in B} E_J(t)$$

$$\sum_{J \in B} \Pi_J(t) = \sum_{J \in B} E_J(t) \text{ if } \tau_b > 0 \tag{2.7'}$$

$$\tau_b(t) \geq 0.$$

Each region is in exactly one trading bloc. The most frequent number of trading blocs in the cases considered in this book is one—the entire world.

The next equation is the definition of per capita consumption:

$$c_J(t) = C_J(t)/L_J(t). \tag{2.8}$$

The evolution of the capital stock is given by

$$K_J(t) = K_J(t-1)(1-\delta_K)^{10} + 10 \times I_J(t-1). \tag{2.9}$$

where $\delta_K$ is the annual rate of depreciation of the capital stock. We assume that capital depreciates at 10 percent per annum. The coefficient of 10 on $I_J(t-1)$ in equation (2.9) reflects the convention that investment is measured at annual rates, while the period in the model is ten years.

The next set of relations involves the supply side of the energy market. The cost of carbon-energy is:

$$c^E_J(t) = q(t) + Markup^E_J(t), \tag{2.10}$$

where $c^E_J(t)$ is the cost per unit of carbon-energy in region $J$, $q(t)$ is the wholesale price of carbon-energy exclusive of the Hotelling rent, and $Markup^E_J(t)$ is a markup on energy costs. The wholesale price, $q(t)$, is assumed to be equalized in different regions. The markup captures regional differences in transportation, distribution costs, and national energy taxes and is assumed to be constant over time. Energy taxes are interpreted as Pigovian taxes that reflect the external costs of energy production and consumption.

Note that the cost of carbon-energy in equation (2.10) does not depend on $\varsigma_J(t)$, the ratio of carbon-energy to carbon services. Carbon-saving technical change has been modeled so that it has no output-enhancing effect. In RICE-99, total factor productivity, $A_J(t)$, increases aggregate productivity, but the role of decarbonization, $\varsigma_J(t)$, is to reduce the ratio of carbon emissions to carbon-energy.

The next equation defines cumulative industrial emissions of carbon:

$$CumC(t) = CumC(t-1) + 10 \times E(t), \tag{2.11}$$

where $CumC(t)$ is the cumulative consumption of carbon-energy at the end of period $t$ and $E(t)$ is world use of carbon-energy in period $t$. $E(t)$ is the sum of carbon-energy use across regions.

The next equation represents the supply curve of carbon-energy:

$$q(t) = \xi_1 + \xi_2 [CumC(t)/CumC^*]^{\xi_3}. \tag{2.12}$$

In equation (2.12), $q(t)$ is the wholesale (supply) price of carbon-energy while $\xi_1$, $\xi_2$, and $\xi_3$ are parameters.[4] $CumC^*$ is a parameter that represents the inflection point beyond which the marginal cost of carbon-energy begins to rise sharply.

### Concentrations, Climate Change, and Damage Equations

The next set of relationships has proven a major challenge because there are no well-established empirical regularities and very little history that can be drawn upon to represent the linkage between economic activity and climate change. As with the economic relationships, it is desirable to use a parsimonious specification so that the theoretical model is transparent and so that the optimization model is empirically tractable. The methodology is drawn from macroeconomics, in which economic behavior is represented by equations that capture the behavior of broad aggregates (such as consumers or investors). The challenge in modeling climate-change economics is that aggregate relationships are needed for optimization approaches like the DICE and RICE models.

The first link is between economic activity and greenhouse-gas emissions. In the DICE-RICE-99 models, greenhouse gases affect climate through their radiative forcing. Of the suite of GHGs, only industrial $CO_2$ is endogenous in the model. The other GHGs (including $CO_2$ arising from land-use changes) are exogenous and projected on the basis of current analysis by the IPCC, the International Institute for Applied Systems Analysis (IIASA), and other scientific groups. Nearly 80 percent of the radiative forcing in 2100 comes from $CO_2$ in RICE-99, and more than 90 percent of cumulative $CO_2$ emissions come from industrial sources, so most of the attention here is devoted to industrial $CO_2$.

---

4. In earlier versions of the revised RICE model, a backstop technology was introduced at a cost of around $500 per ton of carbon. The current RICE-99 and DICE-99 models do not include backstop technologies. Omitting a backstop technology implies that the price of carbon energy can rise to extremely high levels in the future; that also implies that the current Hotelling rent will be high relative to the with backstop model and that emissions in the RICE-99 model are therefore somewhat lower than in a model with a backstop technology. Experiments indicate that the effect of adding a backstop technology is relatively small over the next century and not worth the additional complexity.

In the original DICE model, the accumulation and transportation of emissions were assumed to follow a simple process in which $CO_2$ decayed in the atmosphere at a constant rate. This has been revised in light of inconsistencies with established carbon-cycle modeling.

The new treatment uses a structural approach with a three-reservoir model calibrated to existing carbon-cycle models. The basic idea is that the deep oceans provide a limited, albeit vast, sink for carbon in the long run. In the new specification, we assume that there are three reservoirs for carbon: the atmosphere, a quickly mixing reservoir in the upper oceans and the biosphere, and the deep oceans. Each of the three reservoirs is assumed to be well-mixed in the short run, while the mixing between the upper reservoirs and the deep oceans is assumed to be extremely slow. We assume that $CO_2$ accumulation and transportation can be represented as the following linear three-reservoir model.

$$M_{AT}(t) = 10 \times ET(t-1) + \phi_{11}M_{AT}(t-1) + \phi_{21}M_{UP}(t-1). \tag{2.13a}$$

$$M_{UP}(t) = \phi_{22}M_{UP}(t-1) + \phi_{12}M_{AT}(t-1) + \phi_{32}M_{LO}(t-1). \tag{2.13b}$$

$$M_{LO}(t) = \phi_{33}M_{LO}(t-1) + \phi_{23}M_{UP}(t-1). \tag{2.13c}$$

where $M_{AT}(t)$ is the end-of-period mass of carbon in the atmosphere, $M_{UP}(t)$ is the mass of carbon in the upper reservoir (biosphere, and upper oceans), $ET(t)$ is global $CO_2$ emissions including those arising from land-use changes, and $M_{LO}(t)$ is the mass of carbon in the lower oceans. The coefficient $\phi_{ij}$ is the transfer rate from reservoir $i$ to reservoir $j$ (per period), where $i$ and $j$ = AT, UP, and LO. The calibration of equations (2.13a), (2.13b), and (2.13c) is described in chapter three.

The next step concerns the relationship between the accumulation of GHGs and climate change. This sector uses the same specification as in the original DICE-RICE models because there have been no major developments that would lead to a revision of the underlying approach. Climate modelers have developed a wide variety of approaches for estimating the impact of rising GHGs on climatic variables. On the whole, existing models are much too complex to be included in economic models, particularly ones that are used for optimization. Instead, a small structural model is employed that captures the basic relationship among GHG concentrations, radiative forcings, and the dynamics of climate change.

Accumulations of GHGs lead to global warming through increasing the warming at the surface by increased radiation. The relationship between GHG accumulations and increased radiative forcing, $F(t)$, is derived from empirical measurements and climate models. The relationship is characterized as follows:

$$F(t) = \eta\{\log[M_{AT}(t)/M_{AT}^{PI}]/\log(2)\} + O(t), \tag{2.14}$$

where $M_{AT}(t)$ is the atmospheric concentration of $CO_2$ in billion metric tons of carbon (GtC) and $F(t)$ is the increase in radiative forcing since 1900 in watts per square meter (W/m$^2$), which is the standard measure of radiative forcing. $O(t)$ represents the forcings of other GHGs (CFCs, $CH_4$, $N_2O$, and ozone) and aerosols. These other gases represent a small fraction of the total warming potential; their sources are poorly understood and techniques for preventing their buildup are sketchy today; they are therefore taken as exogenous. The term $M_{AT}^{PI}$ is the preindustrial level of atmospheric concentrations of $CO_2$ (taken to be 596.4 GtC, which is about 280 parts per million).

The list of exogenous components of forcing included in $O(t)$ represents a departure from previous versions of the RICE-DICE models, which considered CFCs to be endogenous and did not include the effects of aerosols. The forcings from non-$CO_2$ GHGs and aerosols are much lower in the current version, reflecting lower anticipated effects of CFCs and the cooling effect of aerosols. These offset slightly higher projections of forcing from methane, nitrous oxide, and tropospheric ozone. All these issues are discussed in detail in the next chapter.

The parameterization of radiative forcing from $CO_2$ in equation (2.14) is not controversial. It relies upon a variety of data on atmospheric concentrations and combines those into a series on radiative forcing as described in the most recent comprehensive IPCC report (IPCC [1996a]). The major assumption for the present modeling is the finding that a doubling of $CO_2$ concentrations would lead to an increase in radiative forcing by 4.1 W/m$^2$.

The next set of equations provides the link between radiative forcing and climate change. Here again, the specification is identical to the original DICE/RICE models. Higher radiative forcings warm the atmospheric layer, which then warms the upper ocean, gradually warming the deep oceans. The lags in the system are primarily due to the thermal inertia of the different layers. The model can be written as follows:

$$T(t) = T(t-1) + \sigma_1\{F(t) - \lambda T(t-1) - \sigma_2[T(t-1) - T_{LO}(t-1)]\}. \tag{2.15a}$$

$$T_{LO}(t) = T_{LO}(t-1) + \sigma_3[T(t-1) - T_{LO}(t-1)]. \tag{2.15b}$$

where $T(t)$ is the increase in the globally and seasonally averaged temperature in the atmosphere and the upper level of the ocean since 1900. $T_{LO}(t)$ is the increase of temperature in the deep oceans. $F(t)$ is the increase in radiative forcing in the atmosphere, $\lambda$ is a feedback parameter, and the $\sigma_i$ are transfer coefficients reflecting the rates of flow and the thermal capacities of the different sinks.

Equations (2.15a) and (2.15b) can be understood as a simple example of the impact of a warming source on a pool of water. Suppose that a heating lamp is turned on (this being the increase in $F(t)$ or radiative forcings). The top part of the pool along with the air at the top are gradually warmed, and the lower part of the pool is gradually warmed as the heat diffuses to the bottom. The lags in the warming of the surface in this simple example are determined by the size of the pool (that is, by its thermal inertia) and by the rate of mixing of the different levels of the pool. This set of equations was fully described for the original DICE model in Nordhaus 1994b.

The next link in the chain is the economic impact of climate change on human and natural systems. Estimating the damages from greenhouse warming has proven extremely elusive. For the purpose of this book, it is assumed that there is a relationship between the damage from greenhouse warming and the extent of warming. More specifically, the relationship between global-temperature increase and income loss is given by:

$$D_J(t) = \theta_{1,J}T(t) + \theta_{2,J}T(t)^2, \tag{2.16}$$

where $D_J(t)$ is the damage from climate change for a region as a fraction of its output net of climate damages and relates the damage to the change in global mean temperature. The damage function is a quadratic function, and the damage relationships are described in chapter 4.

Finally, the damage function can be included into the production function in equation (2.5) using the $\Omega$ coefficient as follows:

$$\Omega_J(t) = 1/[1 + D_J(t)], \tag{2.17}$$

Equations (2.1) through (2.17) form the RICE-99 model that is analyzed in subsequent chapters. Appendix A lists the equations of RICE-99 in a single place. The major variables are summarized in appendix C. The GAMS computer code for the RICE-99 model is listed in appendix D.

## Equilibrium in the Market for Carbon-Energy

In a competitive equilibrium of the model sketched above, the demand for carbon-energy satisfies the following condition:

$$\beta_J \Lambda_J(t) ES_J(t)^{\beta_J-1} = c_J^E(t) + h(t)/\varsigma_J(t) + \tau_J(t)/\varsigma_J(t), \tag{2.18a}$$

where $\Lambda_J(t)$ is a scaling factor that equals $\Omega_J(t) A_J(t) K_J(t)^\gamma L_J(t)^{1-\gamma-\beta_J}$. Rewriting, we obtain:

$$\begin{aligned} E_J(t) &= [1/\varsigma_J(t)]\{[c_J^E(t) + h(t)/\varsigma_J(t) + \tau_J(t)/\varsigma_J(t)]/\beta_J \Lambda_J(t)\}^{[1/(\beta_J-1)]} \\ &= \zeta_J^t[\tau_J(t)]. \end{aligned} \tag{2.18b}$$

The market price includes three terms: the cost of production of carbon-energy, the Hotelling rent representing the effect of current extraction of carbon fuels on future extraction costs, and the carbon tax. Both the carbon tax and the Hotelling rent are applied only to the carbon content of carbon-energy; they are therefore adjusted by the ratio of carbon to carbon-energy in equation (2.18a). Subtracting the regional markup from the market price yields the wholesale price of carbon-energy.

Summing equation (2.18b) across regions in a trading bloc and substituting in (2.7′), we get the equilibrium condition in the market for industrial emissions permits:

$$\sum_{J \in B} \Pi_J(t) \ge \sum_{J \in B} \zeta^t{}_J(\tau_b(t)), \tag{2.19}$$

with the inequality becoming an equality if the carbon tax is greater than zero. $\zeta_J^t[\tau_J(t)]$ is the right-hand side of equation (2.18b) which states that total demand for emissions in a trading bloc cannot exceed the supply.

## Policy in RICE-99

Policymakers (or modelers analyzing policy) can use either carbon taxes or emissions permits as the instrument of policy in RICE-99. In practice, there are many ways to accomplish these indirectly or in combination.

Equation (2.19) says that a policymaker can view either the carbon tax in each trading bloc or the total emissions permits allocated to each trading bloc as a policy variable. If the policymaker specifies the total permits for a trading bloc, then the carbon tax is determined by the necessity to equate demand and supply. If the policymaker specifies the carbon tax, then the total permit allocation of a trading bloc is determined, although the policymaker can choose how to split up the permits among the members of the trading bloc. The user can always satisfy equation (2.19) for any schedule of carbon taxes by simply granting each region permits equal to its emissions from equation (2.18b) at the market or equilibrium carbon tax.

Setting the carbon tax to zero in all regions will produce the reference or baseline case of the model, a projection of what will happen if no government action is taken to slow global warming. In the baseline case, emissions are determined by an unregulated market.

A Pareto-optimal policy—designed as a policy that induces the economically efficient level of emissions—can be achieved by setting the carbon tax in each region equal to the global environmental shadow price of carbon. The environmental shadow price of carbon is the impact through environmental channels of a unit of emissions today on the present value of consumption in all regions in all future periods.

As will be seen in later simulations, policies to slow global warming will have quite different costs and benefits in different regions. Some regions are likely to be more affected by climate change, and the costs of an efficient policy are also likely to be quite asymmetric. The allocation of carbon permits within a trading bloc is a way of influencing the distribution of gains and losses from climate-change policy. Granting a region emissions permit in excess of its emissions will transfer to that region permit revenues that are collected from other regions.[5]

Granting each region a number of permits equal to its emissions will ensure that no transfers occur via permit purchases and sales. A distribution of emissions permits that leads to no redistribution of income among nations is called a *revenue-neutral permit allocation*; this is equivalent to a regime in which countries set harmonized carbon taxes with no transfers among countries.

While the policy choice of the user has been interpreted as a permit-trading arrangement, any combination of taxes and allocated permits that satisfies the constraints above could also be interpreted as a fiscal regime with a given carbon tax and tax revenues. The usual way in which a uniform carbon-tax plan is assumed to work, where regions

---

5. This assumes that the carbon tax is not zero.

harmonize their carbon taxes and each redistributes its revenues to its own citizens in a lump sum fashion, could be implemented in this model by setting carbon taxes equal in all regions and allocating permits in a revenue-neutral fashion.

# 3 Calibration of the Major Sections

## Regional Specification

The data for RICE-99 were collected for thirteen subregions, which were then aggregated into eight regions for modeling purposes. The eight regions were grouped on the basis of either economic or political similarity. The United States and China form two of the eight regions. OECD Europe was treated as a single unit because of the high level of political and economic integration in that region.

The other regions were generally created on the basis of regional or economic similarity. The other high-income group includes Japan, Canada, Australia, and a few other smaller countries. Russia and Eastern Europe includes both Russia and the formerly centrally planned economies of that region, which have extremely high carbon intensities. The significant countries in the middle-income group are Brazil, South Korea, Argentina, Taiwan, Malaysia, and high-income OPEC countries. The lower middle-income region includes Mexico, Turkey, Thailand, South Africa, much of South America, and several populous oil exporters such as Iran. The low-income region, the largest by population, includes South Asia, most of India and Southeast Asia, much of the Asian part of the former Soviet Union, Subsaharan Africa, and a few countries in Latin America.[1]

Tables 3.1 and 3.2 show the composition of the regions as well as the data on $CO_2$ emissions, population, GDP, and GDP growth for each region. Tables 3.3 through 3.5 show calculated data on growth in output per capita, energy intensity, and carbon intensity of the different regions.

---

1. In the tables and text throughout the book, we will occasionally use the following abbreviations: United States—USA, Other High Income—OHI, OECD Europe—Europe, Russia and Eastern Europe—R&EE, Middle Income—MI, Lower Middle Income—LMI, Low Income—LI.

**Table 3.1**
Regional details of the RICE-99 model

| | Industrial CO$_2$ emissions (1,000 tons carbon) 1995 | Gross domestic product (1990 U.S. prices, market exchange rates) | | Population (millions) 1995 | CO$_2$-GDP ratio (tons carbon per $ thousand) 1995 |
| --- | --- | --- | --- | --- | --- |
| | | ($ billions) 1995 | GDP growth rate (percent per year) 1970–95 | | |
| United States | 1,407,257 | 6,176 | 2.6 | 263.12 | 0.23 |
| Other high income | 556,855 | 4,507 | 3.6 | 191.61 | 0.12 |
| Japan | 307,520 | 3,420 | 3.6 | 125.21 | 0.09 |
| Canada | 118,927 | 541 | 3.2 | 29.61 | 0.22 |
| Australia | 79,096 | 295 | 3.1 | 18.05 | 0.27 |
| Singapore | 17,377 | 46 | 8.1 | 2.99 | 0.38 |
| Israel | 12,642 | 66 | 5.0 | 5.52 | 0.19 |
| Hong Kong | 8,459 | 84 | 7.4 | 6.19 | 0.10 |
| New Zealand | 7,489 | 49 | 2.2 | 3.60 | 0.15 |
| Virgin Islands (U.S.) | 3,121 | 2 | NA | 0.10 | 2.01 |
| Guam | 1,129 | .. | NA | .. | NA |
| Aruba | 491 | .. | NA | .. | NA |
| Bahamas | 466 | 3 | NA | 0.28 | 0.14 |
| Bermuda | 124 | 1 | NA | 0.06 | 0.08 |
| British Virgin Islands | 14 | .. | NA | .. | NA |
| Andorra | .. | .. | NA | .. | NA |
| Faeroe Islands | .. | .. | NA | .. | NA |

| | | | | | |
|---|---|---|---|---|---|
| Monaco | .. | .. | NA | .. | NA |
| San Marino | .. | .. | NA | .. | NA |
| *OECD Europe* | 850,839 | 6,892 | 2.4 | 380.85 | 0.12 |
| Germany | 227,920 | 1,787 | 2.3 | 81.87 | 0.13 |
| United Kingdom | 147,964 | 892 | 2.1 | 58.53 | 0.17 |
| Italy | 118,927 | 998 | 2.6 | 57.20 | 0.12 |
| France | 92,818 | 1,189 | 2.5 | 58.06 | 0.08 |
| Spain | 63,211 | 406 | 2.9 | 39.20 | 0.16 |
| Netherlands | 37,093 | 303 | 2.4 | 15.46 | 0.12 |
| Belgium | 28,334 | 189 | 2.3 | 10.15 | 0.15 |
| Greece | 20,820 | 60 | 2.5 | 10.47 | 0.35 |
| Norway | 19,774 | 125 | 3.5 | 4.35 | 0.16 |
| Austria | 16,179 | 165 | 2.7 | 5.11 | 0.10 |
| Denmark | 14,975 | 132 | 2.1 | 5.22 | 0.11 |
| Portugal | 14,172 | 58 | 3.3 | 9.93 | 0.24 |
| Finland | 13,923 | 107 | 2.4 | 5.11 | 0.13 |
| Sweden | 12,170 | 195 | 1.6 | 8.83 | 0.06 |
| Switzerland | 10,604 | 213 | 1.4 | 7.04 | 0.05 |
| Ireland | 8,798 | 53 | 4.2 | 3.59 | 0.16 |
| Luxembourg | 2,528 | 13 | NA | 0.41 | 0.20 |
| Iceland | 492 | 6 | NA | 0.27 | 0.08 |
| Greenland | 137 | 1 | NA | 0.06 | NA |
| Liechtenstein | .. | .. | NA | .. | NA |
| *Russia and Eastern Europe* | 863,849 | 1,095 | 1.6 | 535.09 | 0.79 |
| Russia | 496,182 | 334 | 1.2 | 148.20 | 1.48 |

**Table 3.1** (continued)

| | Industrial $CO_2$ emissions (1,000 tons carbon) 1995 | Gross domestic product (1990 U.S. prices, market exchange rates) | | Population (millions) 1995 | $CO_2$-GDP ratio (tons carbon per $ thousand) 1995 |
| --- | --- | --- | --- | --- | --- |
| | | ($ billions) 1995 | GDP growth rate (percent per year) 1970–95 | | |
| Eastern Europe | 367,667 | 380 | 2.8 | 193 | 0.95 |
| Ukraine | 119,599 | 34 | 1.0 | 51.55 | 3.55 |
| Poland | 92,818 | 74 | NA | 38.61 | 1.25 |
| Romania | 33,049 | 35 | NA | 22.69 | 0.95 |
| Czech Republic | 30,581 | 37 | 10.9 | 10.33 | 0.83 |
| Belarus | 16,185 | 20 | 1.2 | 10.34 | 0.81 |
| Bulgaria | 15,474 | 25 | NA | 8.41 | 0.62 |
| Hungary | 15,250 | 27 | 2.2 | 10.23 | 0.56 |
| Slovakia | 10,381 | 19 | 10.7 | 5.37 | 0.56 |
| Serbia and Montenegro | 9,026 | 60 | NA | 10.54 | 0.15 |
| Croatia | 4,644 | 9 | NA | 4.78 | 0.49 |
| Estonia | 4,488 | 4 | 0.7 | 1.48 | 1.03 |
| Lithuania | 4,043 | 8 | 1.0 | 3.72 | 0.51 |
| Slovenia | 3,197 | 8 | NA | 1.99 | 0.41 |
| Moldova | 2,952 | 2 | −0.9 | 4.34 | 1.54 |
| Macedonia, F.Y.R. | 2,934 | 4 | NA | 2.16 | 0.69 |
| Latvia | 2,543 | 6 | 0.6 | 2.52 | 0.46 |
| Bosnia and Hercegovina | 503 | 9 | NA | 4.38 | 0.06 |

| | | | | |
|---|---|---|---|---|
| *Middle income* | 427,153 | 1,372 | 4.7 | 323.67 | 0.31 |
| Korea, Rep. | 101,963 | 288 | 8.8 | 44.85 | 0.35 |
| Brazil | 68,012 | 370 | 4.5 | 159.22 | 0.18 |
| Taiwan | 46,720 | 195 | NA | 21.30 | 0.24 |
| Argentina | 35,334 | 149 | 1.8 | 34.67 | 0.24 |
| Malaysia | 29,095 | 71 | 7.3 | 20.14 | 0.41 |
| Trinidad and Tobago | 4,670 | 6 | NA | 1.29 | 0.84 |
| Puerto Rico | 4,240 | 36 | NA | 3.72 | 0.12 |
| Netherlands Antilles | 1,762 | : | NA | : | NA |
| Cyprus | 1,413 | 7 | NA | 0.73 | 0.21 |
| Gabon | 967 | 6 | NA | 1.10 | 0.17 |
| Suriname | 587 | 2 | NA | 0.43 | 0.36 |
| Martinique | 556 | : | NA | : | NA |
| Malta | 471 | 3 | NA | 0.37 | 0.16 |
| New Caledonia | 468 | : | NA | : | NA |
| Reunion | 424 | : | NA | : | NA |
| Macao | 336 | 4 | NA | 1.97 | 0.08 |
| Barbados | 225 | 0 | NA | 1.53 | 0.75 |
| French Polynesia | 153 | : | NA | : | NA |
| Antigua and Barbuda | 88 | 0 | NA | 0.07 | 0.20 |
| Gibraltar | 62 | : | NA | : | NA |
| St. Lucia | 52 | 0 | NA | 0.16 | 0.11 |
| Seychelles | 44 | 0 | NA | 0.08 | 0.11 |
| Nauru | 38 | : | NA | : | NA |
| St. Kitts and Nevis | 26 | 0 | NA | 0.04 | 0.14 |
| St. Pierre and Miquelon | 19 | : | NA | : | NA |

**Table 3.1** (continued)

| | Industrial CO$_2$ emissions (1,000 tons carbon) 1995 | Gross domestic product (1990 U.S. prices, market exchange rates) | | Population (millions) 1995 | CO$_2$-GDP ratio (tons carbon per $ thousand) 1995 |
| | | ($ billions) 1995 | GDP growth rate (percent per year) 1970–95 | | |
|---|---|---|---|---|---|
| Montserrat | 12 | : | NA | : | NA |
| Turks and Caicos islands | 0 | : | NA | : | NA |
| Isle of Man | : | : | NA | : | NA |
| Northern Mariana Islands | : | : | NA | : | NA |
| Anguilla | : | : | NA | : | NA |
| High-income OPEC | 129,416 | 234 | 3.7 | 32.03 | 0.55 |
| United Arab Emirates | 18,642 | 37 | NA | 2.46 | 0.50 |
| Qatar | 7,920 | 7 | NA | 0.64 | 1.07 |
| Kuwait | 13,297 | 32 | NA | 1.55 | 0.41 |
| Saudi Arabia | 69,392 | 108 | 3.7 | 18.98 | 0.64 |
| Libya | 10,754 | 27 | NA | 5.41 | 0.40 |
| Oman | 3,116 | 14 | NA | 2.14 | 0.22 |
| Bahrain | 4,048 | 5 | NA | 0.58 | 0.77 |
| Brunei | 2,247 | 4 | NA | 0.29 | 0.64 |
| *Lower middle income* | 560,578 | 1,156 | 3.7 | 571.42 | 0.43 |
| Mexico | 97,662 | 179 | 3.4 | 91.83 | 0.54 |
| South Africa | 83,462 | 102 | 2.1 | 41.16 | 0.82 |
| Iran, Islamic Rep. | 71,987 | 211 | NA | 64.12 | 0.34 |

| | | | | |
|---|---|---|---|---|
| Venezuela | 49,193 | 65 | 2.0 | 21.67 | 0.76 |
| Turkey | 47,773 | 129 | 4.3 | 61.06 | 0.37 |
| Thailand | 47,773 | 122 | 7.5 | 58.24 | 0.39 |
| Kazakhstan | 37,093 | 18 | 1.6 | 16.61 | 2.04 |
| Algeria | 24,909 | 76 | 3.4 | 27.96 | 0.33 |
| Colombia | 18,429 | 57 | 4.5 | 36.81 | 0.32 |
| Syrian Arab Rep. | 12,561 | 20 | 6.2 | 14.11 | 0.62 |
| Chile | 12,037 | 16 | 5.2 | 14.23 | 0.75 |
| Peru | 8,341 | 28 | 2.2 | 23.82 | 0.30 |
| Morocco | 7,995 | 26 | 3.9 | 26.56 | 0.31 |
| Cuba | 7,933 | 23 | NA | 11.01 | 0.35 |
| Turkmenistan | 7,733 | 1 | 3.6 | 4.51 | 5.96 |
| Ecuador | 6,177 | 7 | 4.5 | 11.48 | 0.83 |
| Tunisia | 4,178 | 15 | 5.1 | 8.96 | 0.29 |
| Dominican Rep. | 3,212 | 7 | 4.5 | 7.82 | 0.43 |
| Jamaica | 2,470 | 4 | NA | 2.52 | 0.59 |
| Panama | 1,882 | 8 | NA | 2.63 | 0.24 |
| Uruguay | 1,468 | 10 | 1.8 | 3.18 | 0.15 |
| Costa Rica | 1,428 | 7 | 4.1 | 3.40 | 0.20 |
| El Salvador | 1,416 | 7 | 1.9 | 5.62 | 0.22 |
| Paraguay | 1,036 | 6 | 5.2 | 4.83 | 0.18 |
| Papua New Guinea | 677 | 5 | 3.1 | 4.30 | 0.13 |
| Guadeloupe | 416 | : | NA | : | NA |
| Mauritius | 407 | 3 | NA | 1.12 | 0.13 |
| French Guiana | 238 | : | NA | : | NA |
| Fiji | 201 | 2 | NA | 0.79 | 0.11 |

**Table 3.1** (continued)

| | Industrial CO₂ emissions (1,000 tons carbon) 1995 | Gross domestic product (1990 U.S. prices, market exchange rates) | | Population (millions) 1995 | CO₂-GDP ratio (tons carbon per $ thousand) 1995 |
| | | ($ billions) 1995 | GDP growth rate (percent per year) 1970–95 | | |
|---|---|---|---|---|---|
| Belize | 113 | 1 | NA | 0.22 | 0.21 |
| Cayman Islands | 84 | : | NA | : | NA |
| American Samoa | 75 | : | NA | : | NA |
| Pacific Islands | 65 | : | NA | : | NA |
| Grenada | 46 | 0 | NA | 0.10 | 0.21 |
| St. Vincent and the Grenadines | 34 | 0 | NA | 0.11 | 0.15 |
| Tonga | 28 | 0 | NA | 0.10 | 0.28 |
| Dominica | 22 | 0 | NA | 0.07 | 0.13 |
| Vanuatu | 17 | 0 | NA | 0.17 | 0.11 |
| Cook Islands | 6 | : | NA | : | NA |
| Niue | 1 | : | NA | : | NA |
| Namibia | : | : | NA | : | NA |
| Micronesia | : | : | NA | : | NA |
| Marshall Islands | : | : | NA | : | NA |
| Wallis and Futuna | : | : | NA | : | NA |
| China | 871,311 | 654 | 8.5 | 1,200.24 | 1.33 |
| Low income | 620,793 | 1,216 | 3.4 | 2,377.02 | 0.51 |
| India | 248,017 | 447 | 4.4 | 929.36 | 0.55 |

| | | | | |
|---|---|---|---|---|
| Indonesia | 80,822 | 158 | 7.1 | 193.28 | 0.51 |
| Korea, Dem. Rep. | 70,138 | 15 | NA | 23.87 | 4.82 |
| Iraq | 27,020 | 12 | NA | 20.10 | 2.33 |
| Uzbekistan | 26,986 | 15 | 3.3 | 22.77 | 1.77 |
| Egypt, Arab Rep. | 25,023 | 48 | 5.4 | 57.80 | 0.53 |
| Pakistan | 23,296 | 56 | 5.3 | 129.91 | 0.42 |
| Philippines | 16,692 | 49 | 3.4 | 68.60 | 0.34 |
| Azerbaijan | 11,620 | 3 | -0.2 | 7.51 | 3.52 |
| Viet Nam | 8,654 | 68 | NA | 73.48 | 0.13 |
| Bangladesh | 5,713 | 27 | 3.3 | 57.80 | 0.21 |
| Yemen | 3,933 | 11 | NA | 15.27 | 0.37 |
| Lebanon | 3,641 | 6 | NA | 4.01 | 0.57 |
| Jordan | 3,632 | 9 | NA | 4.21 | 0.40 |
| Bolivia | 2,859 | 7 | 2.5 | 57.80 | 0.43 |
| Mongolia | 2,308 | 4 | NA | 2.46 | 0.55 |
| Georgia | 2,114 | 3 | -3.4 | 5.40 | 0.80 |
| Guatemala | 1,962 | 11 | 3.4 | 10.62 | 0.18 |
| Myanmar | 1,919 | 15 | NA | 45.11 | 0.13 |
| Sri Lanka | 1,607 | 10 | 4.5 | 18.11 | 0.15 |
| Kyrgyzstan | 1,491 | 1 | 2.4 | 4.52 | 1.18 |
| Honduras | 1,052 | 6 | 3.8 | 5.92 | 0.17 |
| Tajikistan | 1,021 | 2 | 3.4 | 5.84 | 0.61 |
| Armenia | 996 | 1 | -0.1 | 3.76 | 0.83 |
| Nicaragua | 737 | 4 | -0.2 | 4.38 | 0.18 |
| Albania | 504 | 3 | NA | 3.26 | 0.15 |
| Nepal | 418 | 5 | 3.7 | 21.46 | 0.08 |

**Table 3.1** (continued)

| | Industrial CO$_2$ emissions (1,000 tons carbon) 1995 | Gross domestic product (1990 U.S. prices, market exchange rates) | | Population (millions) 1995 | CO$_2$-GDP ratio (tons carbon per $ thousand) 1995 |
| | | ($ billions) 1995 | GDP growth rate (percent per year) 1970–95 | | |
| --- | --- | --- | --- | --- | --- |
| Afghanistan | 338 | 14 | NA | 23.48 | 0.02 |
| Guyana | 255 | 1 | NA | 0.83 | 0.50 |
| Haiti | 174 | 2 | 0.4 | 7.17 | 0.09 |
| Cambodia | 136 | 2 | NA | 10.02 | 0.09 |
| Lao, PDR | 84 | 2 | NA | 4.88 | 0.04 |
| Bhutan | 65 | 0 | NA | 0.70 | 0.14 |
| Western Sahara | 57 | .. | NA | .. | NA |
| Maldives | 50 | 0 | NA | 0.25 | 0.25 |
| Solomon Islands | 44 | 0 | NA | 0.38 | 0.16 |
| Western Samoa | 36 | 0 | NA | 0.17 | 0.36 |
| Sao Tome and Principe | 21 | 0 | NA | 0.13 | 0.29 |
| Kiribati | 6 | 0 | NA | 0.08 | 0.17 |
| West Bank | .. | .. | NA | .. | NA |
| Gaza Strip | .. | .. | NA | .. | NA |
| Tuvalu | .. | .. | NA | .. | NA |
| Tokelau | .. | .. | NA | .. | NA |
| Africa | 45,352 | 199 | 2.7 | 532 | 0.23 |
| Swaziland | 124 | 1 | NA | 0.90 | 0.13 |
| Lesotho | .. | .. | NA | .. | NA |

| Country | | | | |
|---|---|---|---|---|
| Nigeria | 24,759 | 45 | 2.9 | 111.27 | 0.55 |
| Cote d'Ivoire | 2,828 | 12 | 2.6 | 13.98 | 0.24 |
| Sudan | 955 | 14 | NA | 26.71 | 0.07 |
| Kenya | 1,824 | 11 | 5.2 | 26.69 | 0.16 |
| Angola | 1,256 | 8 | NA | 10.77 | 0.16 |
| Botswana | 612 | 3 | NA | 1.46 | 0.20 |
| Congo | 346 | 3 | NA | 2.63 | 0.13 |
| Zaire | 573 | 5 | NA | 43.85 | 0.11 |
| Zimbabwe | 2,657 | 8 | 2.9 | 11.01 | 0.35 |
| Ethiopia | 962 | 10 | NA | 56.40 | 0.10 |
| Senegal | 836 | 6 | 2.5 | 8.47 | 0.13 |
| Ghana | 1,104 | 8 | 1.9 | 17.08 | 0.14 |
| Zambia | 656 | 3 | 0.9 | 8.98 | 0.25 |
| Madagascar | 307 | 3 | 0.5 | 13.65 | 0.10 |
| Guinea | 295 | 3 | NA | 6.59 | 0.10 |
| Cameroon | 1,131 | 11 | 3.3 | 13.29 | 0.10 |
| Uganda | 285 | 12 | NA | 19.17 | 0.02 |
| Niger | 305 | 3 | 0.3 | 9.03 | 0.11 |
| Mali | 127 | 3 | 3.0 | 9.79 | 0.04 |
| Rwanda | 134 | 1 | 0.8 | 6.40 | 0.09 |
| Malawi | 198 | 2 | 3.7 | 9.76 | 0.12 |
| Benin | 173 | 2 | NA | 5.48 | 0.08 |
| Somalia | 3 | 1 | NA | 9.49 | 0.00 |
| Togo | 203 | 2 | 2.0 | 4.09 | 0.13 |
| Tanzania | 666 | 5 | NA | 29.65 | 0.13 |
| Burkina Faso | 261 | 3 | 3.6 | 10.38 | 0.09 |

**Table 3.1** (continued)

| | Industrial $CO_2$ emissions (1,000 tons carbon) 1995 | Gross domestic product (1990 U.S. prices, market exchange rates) ($ billions) 1995 | GDP growth rate (percent per year) 1970–95 | Population (millions) 1995 | $CO_2$-GDP ratio (tons carbon per $ thousand) 1995 |
|---|---|---|---|---|---|
| Mozambique | 271 | 2 | NA | 16.17 | 0.11 |
| Central African Rep. | 64 | 1 | 1.4 | 3.28 | 0.05 |
| Chad | 26 | 1 | 1.9 | 6.45 | 0.02 |
| Burundi | 58 | 1 | 2.8 | 6.26 | 0.04 |
| Mauritania | 837 | 1 | NA | 2.27 | 0.65 |
| Liberia | 87 | 1 | NA | 2.73 | 0.07 |
| Sierra Leone | 121 | 1 | 0.8 | 4.20 | 0.15 |
| Djibouti | 101 | 0 | NA | 0.60 | 0.24 |
| Gambia, The | 59 | 0 | NA | 1.11 | 0.19 |
| Cape Verde | 31 | 0 | NA | 0.38 | 0.08 |
| Comoros | 18 | 0 | NA | 0.49 | 0.08 |
| Guinea-Bissau | 63 | 0 | NA | 1.07 | 0.24 |
| Equatorial Guinea | 36 | 0 | NA | 0.40 | 0.12 |

Sources to tables 3.1–3.5: Output and population data are from *World Development Indicators, 1998*, CD-ROM. Energy data for 1995 were from United Nations, *1995 Energy Statistics Yearbook*, New York 1997. Energy data for 1970 were from United Nations Dept. Of Economic and Social Affairs Statistical Office, *World Energy Supplies, 1970–73*, New York, 1975. Energy data for other years were from the World Bank, *World Development Indicators 1998*. Data on carbon dioxide emissions were from the web page of the Carbon Dioxide Information Analysis Center.

**Table 3.2**
Major regional aggregates in RICE-99 regions

| | Industrial $CO_2$ emissions (million metric tons, carbon weight) 1995 | Gross domestic product (1990 U.S. prices, market exchange rates) | | | Population (millions) 1995 | $CO_2$-GDP ratio (tons per $1,000) 1995 |
| | | (billions of $) 1995 | Growth rate of real GDP (% per year) 1970–95 | GDP per capita 1995 | | |
|---|---|---|---|---|---|---|
| United States | 1,407.3 | 6,176 | 2.6 | 23,472 | 263.1 | 0.23 |
| Other high income | 556.9 | 4,507 | 3.6 | 23,522 | 191.6 | 0.12 |
| OECD Europe | 850.8 | 6,891 | 2.4 | 18,094 | 380.9 | 0.12 |
| Russia and Eastern Europe | 863.8 | 693 | 1.6 | 2,098 | 341.6 | 1.24 |
| Middle income | 427.2 | 1,372 | 4.7 | 4,239 | 323.7 | 0.31 |
| Lower-middle income | 560.6 | 1,156 | 3.7 | 2,023 | 571.4 | 0.48 |
| China | 871.3 | 654 | 8.5 | 545 | 1,200.3 | 1.33 |
| Low income | 620.8 | 1,216 | 3.4 | 512 | 2,377.0 | 0.51 |
| World | 6,158.7 | 22,665 | 3.0 | 4,020 | 5,649.7 | 0.51 |

**Table 3.3**
Growth rates of per capita GDP: Regional averages (percent per year, annual average)

|                          | 1970–75 | 1975–80 | 1980–85 | 1985–90 | 1990–95 | 1970–95 |
|--------------------------|---------|---------|---------|---------|---------|---------|
| United States            | 1.10    | 2.08    | 1.60    | 1.83    | 1.29    | 1.58    |
| Other high income        | 2.95    | 3.16    | 2.42    | 3.53    | 0.96    | 2.60    |
| OECD Europe              | 2.32    | 2.70    | 1.37    | 2.74    | 0.95    | 2.01    |
| Russia and Eastern Europe| 5.20    | 4.20    | 2.75    | 1.08    | −7.04   | 1.14    |
| Middle income            | 4.51    | 4.14    | −1.91   | 1.55    | 3.04    | 2.24    |
| Lower middle income      | 2.46    | 2.31    | 0.13    | 0.92    | 0.75    | 1.31    |
| China                    | 3.34    | 4.89    | 8.72    | 6.30    | 11.08   | 6.83    |
| Low income               | 1.87    | 2.53    | −0.37   | 1.25    | 0.07    | 1.06    |
| World                    | 2.18    | 2.96    | 0.37    | 1.91    | 0.62    | 1.60    |

**Table 3.4**
Growth rates of commercial energy/GDP ratio: Regional aggregates (percent per year, annual average)

|                          | 1970–75 | 1975–80 | 1980–85 | 1985–90 | 1990–95 | 1970–95 |
|--------------------------|---------|---------|---------|---------|---------|---------|
| United States            | 0.03    | −1.32   | −2.86   | −1.10   | −1.67   | −1.39   |
| Other high income        | 1.90    | −1.51   | −2.48   | −1.00   | −1.82   | −0.99   |
| OECD Europe              | 1.43    | −0.81   | −1.19   | −1.74   | −2.95   | −1.06   |
| Russia and Eastern Europe| −2.76   | −2.01   | 1.75    | −2.26   | 0.74    | −0.93   |
| Middle income            | 3.12    | 1.77    | 3.46    | 1.65    | −1.16   | 1.75    |
| Lower middle income      | 9.70    | −0.57   | 3.97    | −1.71   | −2.35   | 1.71    |
| China                    | −1.62   | −0.62   | −5.12   | −2.85   | −8.15   | −3.71   |
| Low income               | −3.83   | 1.16    | 4.95    | 1.33    | −0.20   | 0.64    |
| World                    | 0.78    | −0.69   | −0.02   | −1.39   | −2.38   | −0.75   |

Table 3.5
Growth rates of $CO_2$-GDP ratio: Regional averages (percent per year, annual average)

|  | 1970–75 | 1975–80 | 1980–85 | 1985–90 | 1990–95 | 1970–95 |
|---|---|---|---|---|---|---|
| United States | −1.88 | −1.75 | −3.23 | −0.70 | −1.41 | −1.80 |
| Other high income | −1.44 | −1.94 | −3.15 | −1.61 | −0.30 | −1.70 |
| OECD Europe | −2.03 | −1.32 | −3.18 | −2.68 | −1.94 | −2.23 |
| Russia and Eastern Europe | −1.87 | −2.09 | −1.64 | −2.71 | 2.47 | −1.19 |
| Middle income | −1.63 | 2.10 | 1.85 | 0.97 | 2.48 | 1.14 |
| Lower middle income | 0.10 | −0.05 | 0.25 | −0.43 | −0.51 | −0.13 |
| China | 2.39 | −0.99 | −4.09 | −3.43 | −5.93 | −2.45 |
| Low income | 2.64 | 0.83 | 0.57 | 0.73 | 1.54 | 1.26 |
| World | −0.65 | −1.02 | −1.47 | −1.45 | −0.77 | −1.07 |

Figure 3.1 shows the estimated $CO_2$-output ratios for the thirteen subregions.

## Calibration of Production Function

### Energy-Production Module

RICE-99 uses a new approach to the production structure relative to both the original models and to existing climate-change models. The major changes are RICE-99 introduces a new concept and technique for incorporating energy use by defining an aggregate called "carbon-energy" and production is revised to correspond to a more traditional economic approach to modeling production and inputs choices. The new approach is described in this section and the calibration of the production function is described in the next section.

Output in each region is assumed to be produced by capital, labor, and carbon-energy in a Cobb-Douglas framework (see equation 2.5 in chapter 2). Carbon-energy can be described as the energy services derived from carbon fuels (i.e., fossil fuels). In this approach, $CO_2$ emissions are a joint product of carbon-based energy consumption. Carbon-saving technological change is modeled as reducing the ratio of carbon-energy consumption or carbon emissions to energy services (or the amount of carbon emissions per unit output at constant input prices). Carbon-energy is supplied according to a supply function in

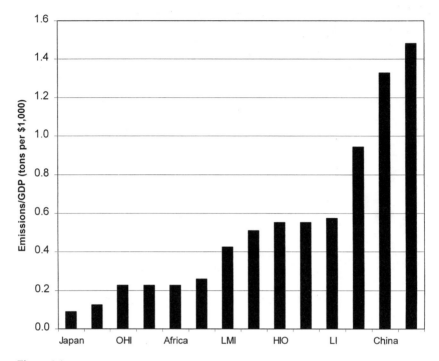

**Figure 3.1**
Industrial $CO_2$-output ratios for thirteen RICE subregions, 1995
For key to regions, see table 3.1.

which the marginal cost increases as carbon fuels are exhausted (this sector is discussed in the fourth section of this chapter). For more detail, see chapter 2, the second section.

The purpose of the aggregation is to simplify the model by having a single energy input and treating all noncarbon fuels as combinations of capital and labor. The advantage of this approach is that it greatly simplifies the enormous complexities of interfuel substitution. The disadvantage is that this approach may lose some of the fine detail of the interfuel substitution relationships, particularly for high carbon taxes. It is important to ensure that the aggregate model has the same behavior over time and with respect to carbon limitations or carbon taxes as would a more complete disaggregated model. To ensure consistency with more complete models, we parameterize the production function so that the response of industrial carbon emissions to a given increase in carbon taxes is the same as the response in a disaggregated model in which emissions are modeled as the sum of emissions across energy sources.

The convention of measuring output, consumption and investment in 1990 U.S. dollars at market exchange rates is followed here. All dollar values in the text, tables, and graphs represent 1990 U.S. dollars. Prices can be converted into prices for the year 2000 (using the U.S. GDP deflator) by multiplying by 1.25.

Although it is common practice to use output measured at international or purchasing-power parity (PPP) exchange rates, this is inappropriate in the current context for three reasons. First, since historical output data at market exchange rates are more readily available than at PPP rates, we rely on these data to make projections about future growth in output and carbon intensity. For the output levels projected to be consistent with projected carbon intensity growth rates, we define them as output at market exchange rates. Second, in the context of optimizing a country's consumption path, it should use its internal prices rather than the world average price level. Third, international trading in carbon emissions permits will take place at market exchange rates, so output needs to be measured in market exchange rates for consistency in measurement between trade flows and economic production as well as between the marginal cost of carbon abatement and the international carbon permit price. If users would like to convert the data to PPP income levels, the levels of output can of course be scaled by a factor to represent living standards at a particular time, but this has little substantive effect on the results.

## Matching Initial Period to the Data

Recall the production function from equation (2.5) in the last chapter. In the first period, we define carbon services to be equal to carbon-energy use [that is, $\varsigma_J(0) = 1$]. This gives the following results:

$$Q_J(0) = \Omega_J(0)\{A_J(0)K_J(0)^\gamma L_J(0)^{1-\beta_J-\gamma} E_J(0)^{\beta_J} - c_J^E(0)E_J(0)\}. \tag{3.1a}$$

$$c_J^E(0) = q(0) + Markup_J^E. \tag{3.1b}$$

$\Omega_J(0)$ and $q(0)$ are determined in a fashion that will be described below. The capital-elasticity coefficient $\gamma$ is assumed to be 0.3 on the basis of standard production-function studies. $L_J(0)$ is the initial population level, taken from the data. $E_J(0)$ is determined endogenously by market participants.

This leaves the parameters $A_J(0)$, $K_J(0)$, $\beta_J$, and $Markup_J^E$ open. These parameters are calibrated so that the model matches empirical observation in four key aspects: GDP, industrial emissions, interest rates, and the effect of a carbon tax on emissions. That is, the parameter values are determined by finding for each region the combination that solves the following four equations in the base case when $t_j(0) = 0$:[2]

First period output = Output from historical data.                    (3.2a)

First period industrial emissions = Industrial emissions
                                                   from historical data.        (3.2b)

First period interest rates = Target value.                           (3.2c)

Change in carbon emissions = Calculated from a
from \$50 carbon tax in            disaggregated energy model.    (3.2d)
first period

Equation (3.2a) sets the right-hand side of (3.1a) equal to actual output for the region, taken from the historical data. Equation (3.2b) sets the right-hand side of (2.18b) equal to actual industrial emissions from the data, assuming $h(0) = 0$.[3]

The interest rate on capital in the first period is its net marginal product. In RICE-99, this is given by its contribution to output and to the next period's capital stock. Equation (3.2c) sets this equal to a target value reflecting historical data and actual differences across regions.

The most important calibration for policy purposes is determining the impact of carbon-emissions constraints. Using equation (3.2d), the parameters are set to equate the model's emissions reductions to the target for a \$50 carbon tax. From (2.18b) the decrease in emissions can be calculated due to the imposition of a \$50 carbon tax. This gives the right-hand side of (3.2d). To get the left-hand side of (3.2d), a disaggregated energy sector model is constructed, where carbon emissions are given by:

$$E(0) = \Sigma X_i(0)\gamma_i. \tag{3.3}$$

$X_i$ is the consumption of energy source $i$ and $\gamma_i$ is $CO_2$ emissions per unit of consumption for energy source $i$. Assume that the demand for each fossil fuel takes the form

$$X_i(0) = \omega_i(0)\{P_i(0)/[P_i(0) + \tau(0)\gamma_i]\}^{\eta_i}, \tag{3.4}$$

---

2. For the fourth equation, we measure the change in emissions assuming the initial carbon tax is 0.
3. The Hotelling rent is calculated to be \$0.39 per ton carbon energy in the first period in the RICE base case, which is essentially zero as compared with $q(0) = \$113$ per ton.

where $\omega_i(0)$ is the consumption of energy source $i$ in the first period, $P_i(0)$ is the price of energy source $i$ in the first period, and $\eta_i$ is the price-elasticity of demand for energy source $i$. Using (3.3) and (3.4), we calculate the change in emissions due to the imposition of a $50 carbon tax, which is then used on the left-hand side of (3.2d).[4]

*Data Sources*

**Population.** Data for the initial population level were obtained from *UN Monthly Bulletin of Statistics, July 1996*.

**GDP.** Data on output were taken from *UN Monthly Bulletin of Statistics, July 1996*.

**Industrial carbon-dioxide emissions.** Data on total industrial carbon emissions were obtained from CDIAC (Carbon Dioxide Information Analysis Center) of Oak Ridge National Labs, U.S. Department of Energy.

**Energy consumption.** The different energy sources $(X_i)$ are nonelectric coal consumption, nonelectric natural gas consumption, electricity consumption, and consumption of petroleum products. Electricity consumption data were taken from *International Energy Annual 1996*, published by the Energy Information Administration (EIA) of the U.S. Department of Energy. Fossil fuel shares for electricity were derived using EIA's World Energy Projection System, 1997 version. Data for total coal and natural gas consumption were taken from *International Energy Annual 1996*, and nonelectric coal and natural gas consumption were calculated as the difference between electricity consumption and total consumption. Petroleum products consumption data were drawn from *International Energy Annual 1995*.

**Energy prices.** Data on electricity prices, petroleum product prices, coal prices, and natural gas prices were obtained from the EIA home

---

4. To get the left-hand side of equation (3.2d) for a given region, we first delete from the data set all countries for which we do not have complete data. We then calculate the change in emissions from each country due to the imposition of the $50 carbon tax, using (3.3) and (3.4). We then sum these up and multiply by the ratio of total industrial carbon emissions in the region, including the countries dropped from the calculation, to the sum of the left-hand side of (3.3) across countries in the restricted data set.

page; *Energy Prices and Taxes*, various issues (published by the International Energy Agency of the OECD); World Bank technical paper number 248, *A Survey of Asia's Energy Prices*; and *International Energy Annual*, various issues.

**Price elasticities.**   After a review of the literature on energy demand, price elasticities of demand for all energy demand components with respect to retail prices were assumed to be $-0.7$ in OECD regions (U.S., OHI, and OECD Europe) and $-0.84$ in the rest of the world.

**Carbon emission factors.**   Carbon coefficients for individual fossil fuels and petroleum products were taken from a variety of publications of the Department of Energy, Energy Information Agency.[5] The carbon coefficient for electricity is the sum of the carbon coefficients of individual fossil fuels weighted by their fuel share in electricity consumption, adjusted for the efficiency of conversion of the fossil fuel into electricity.

### Exogenous Trend Parameters

The previous section described calibration of the model to initial-period data. This section describes calibration of the parameters that drive the trends in GDP and carbon emissions. These parameters govern population growth, total factor productivity growth, and the ratio of carbon services to carbon-energy.

### *Details of the Calibration*

**Empirical estimates for population.**   As described above, the RICE model uses an exponential smoothing model of population growth. In this approach, the model is fitted exactly to three points of the population projection trajectory: the initial population level, the asymptotic level of population, and the initial rate of population growth. For intermediate population levels and growth rates, the technique leads to small approximation errors. The approach is particularly useful for updating the model because of the small number of parameters needed for specification. Note that for two regions whose populations are projected to decline—OECD Europe and Japan—we directly input data for the first few periods and start the exponential model only when projected population starts to decline.

---

5. See EIA 1996.

**Data sources for population.** For initial population, see the previous section. For the first four periods for OECD Europe and Japan, we use the UN population projections for 1995, 2005, 2015, and 2025 from *UN World Population Prospects, 1994 Revision*. For the stationary (asymptotic) population level, we use the World Bank's estimates from *World Bank Population Projections, 1994–5*. The initial rate of population growth is calculated from *World Bank Population Projections, 1994–5*. For some regions, it was modified to improve the match with the World Bank's projections for 2050.

**Estimates of long-run output growth.** There are major uncertainties about the long-run trajectories of economic growth in different regions. Some involve environmental issues, but the most important are likely to be political factors, the presence or absence of wars, and future technological change. Historical growth rates of output per capita for the eight regions in RICE-99 are shown in table 3.3.

One set of information that informed these projections for economic growth was an informal survey of ten economists and economic historians who were asked their views about long-run growth trends. The major assumptions underlying these projections are that the growth rate of output per capita will slow in the twenty-first century in the developed regions and that developing regions will grow at rates that produce partial convergence of output per capita by the end of the next century.

A comparison of the RICE-99 assumptions with those of Angus Maddison (1998a) is provided in table 3.6.[6] Our assumptions are generally in line with those in the Maddison study, with the world weighted average growth rate in RICE-99 approximately 0.16 percent per year higher for the next two decades. The major differences between the two projections are that we are more optimistic about Africa and Latin America and less optimistic about China.

The long-run growth rates, historical rates, and levels of per capita GDP for the different regions RICE-99 are shown in table 3.7.

**Model calibration for long-run output growth.** To model the trend of long-run economic growth, RICE-99 assumes an exogenous exponential trend in technological progress, similar to that used for population described above. The choice of the initial level of productivity

---

6. The RICE-99 projections in the tables in this chapter come from the base case. See chapter 2, section five for explanation.

**Table 3.6**
Comparison of RICE-99 with Maddison projections (percent annual average growth rates of per capita GDP)

| Maddison region | RICE region | Historical estimates | | Projections | | |
|---|---|---|---|---|---|---|
| | | RICE 1970–95 | Maddison 1973–95 | RICE 1995–2015 | Maddison 1995–2015 | Difference |
| United States | United States | 1.58 | 1.55 | 1.36 | 1.30 | 0.06 |
| Japan | Other high income | 2.60 | 2.53 | 1.32 | 1.30 | 0.02 |
| Western Europe | OECD Europe | 2.01 | 1.72 | 1.32 | 1.30 | 0.02 |
| Other Europe | Russia and Eastern Europe | 1.14 | 0.48 | 2.63 | 2.68 | −0.05 |
| Other Americas | Middle income | 2.24 | 0.68 | 2.05 | 1.33 | 0.72 |
| Other Americas | Lower middle income | 1.31 | 0.68 | 2.54 | 1.33 | 1.21 |
| China | China | 6.83 | 5.37 | 3.37 | 4.50 | −1.13 |
| India and Africa[1] | Low income | 1.06 | 1.78 | 2.93 | 2.38 | 0.55 |
| Weighted average[1] | | 2.52 | 2.30 | 2.68 | 2.52 | 0.16 |

Source: Maddison 1998a, 1998b, and 1995 and RICE-99 model, base case.
1. Weighted by 1995 population.

**Table 3.7**
Growth in per capita output in RICE-99 regions: Historical rates and projections

| Region | Actual 1970–95 | RICE-99 model calculations | | | | | | Per capita output (1990 U.S. prices) | |
|---|---|---|---|---|---|---|---|---|---|
| | | Growth rate of per capita output, percent per year | | | | | | Actual 1995 | Projected 2100 |
| | | Projected | | | | | | | |
| | | 1995–2005 | 2005–15 | 2015–25 | 1995–2045 | 2045–95 | | | |
| United States | 1.58 | 1.62 | 1.06 | 0.83 | 0.98 | 0.73 | | $22,862 | $53,480 |
| Other high income | 2.60 | 1.56 | 1.05 | 0.83 | 0.97 | 0.75 | | 22,514 | 53,013 |
| OECD Europe | 2.01 | 1.60 | 1.05 | 0.81 | 0.98 | 0.72 | | 18,818 | 43,822 |
| Russia and Eastern Europe | 1.14 | 3.04 | 2.25 | 1.90 | 2.10 | 1.65 | | 2,095 | 13,443 |
| Middle income | 2.24 | 2.37 | 1.74 | 1.46 | 1.63 | 1.31 | | 5,091 | 21,918 |
| Lower-middle income | 1.31 | 2.91 | 2.14 | 1.80 | 2.00 | 1.52 | | 2,053 | 11,729 |
| China | 6.83 | 3.81 | 2.94 | 2.54 | 2.75 | 2.14 | | 486 | 5,442 |
| Low income | 1.06 | 3.39 | 2.46 | 2.07 | 2.30 | 1.77 | | 436 | 3,265 |

Note: The estimates of output levels and growth rates use market exchange rates to convert to 1990 U.S. dollars. The levels of output are calculated to be substantially higher for low-income countries and lower for Japan and Europe if purchasing power parity (PPP) exchange rates are used. Historical growth rates using PPP exchange rates are not available for most countries.

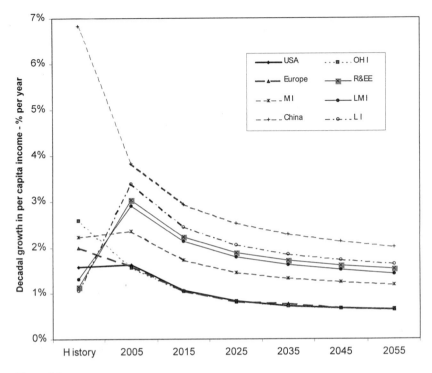

**Figure 3.2**
Growth in per capita output

[the coefficient $A_j(0)$ in the production function] is described above. The initial rate of productivity growth is chosen so that growth between the first and second periods in per capita output matches the assumed rates. The rate of growth of TFP declines at an exponential rate to fit the assumed asymptotic level of output per capita. The historical growth rate of output per capita along with the calculated rates are shown in figure 3.2 and table 3.7.

**Calibration of carbon-saving technological change.** Calibration of the rate of decarbonization (the rate of decline in carbon intensity[7]) is made by adjusting the parameters that govern the ratio of energy services to industrial carbon emissions, $\varsigma_j(t)$ in equation (2.5b). We assume $\varsigma_j(t)$ grows at a declining growth rate in a manner similar to population and total factor productivity. Decarbonization is difficult to project because it depends upon trends in energy sources, on energy-sector

---

7. Carbon intensity is the ratio of industrial carbon emissions to GDP.

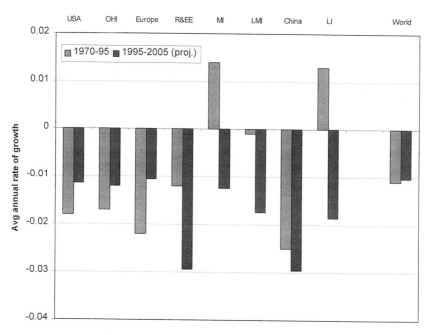

**Figure 3.3**
Rates of growth in $CO_2$ emissions/GDP ratio

technological change, and on policies about energy taxation. These have generally been projected on the basis of historical trends in decarbonization, with adjustments for regions such as Russia and Eastern Europe that have divergent carbon intensities for special historical reasons. Figure 3.3 shows recent trends and near-term projections for the rate of decarbonization. A continuation in the recent trend of decarbonization is projected, with a decline in global carbon intensity of 1.03 percent per year in the 1995–2005 decade compared to a historical decline in the 1970–95 period of 1.07 percent per year. The assumed rate of decarbonization tends toward zero in subsequent decades along with the assumed rate of economy-wide technological change.

The projections developed here can be compared with the systematic survey by Nakicenovic, Grubler, and McDonald [1998]. They provide a range of scenarios, but their scenario B is probably closest to the philosophy expressed in this book. These two studies use methodologies that are completely independent in their underlying construction. Table 3.8 compares the results of the two studies. RICE-99 has somewhat slower output growth and a somewhat slower rate of decarbonization, with the net result being that global industrial emissions

Table 3.8
Comparison of RICE-99 reference case with IIASA scenario B

|  | 2025 | | 2055 | |
|---|---|---|---|---|
|  | IIASA | RICE-99 | IIASA | RICE-99 |
| Population (billions) | 8.3 | 7.6 | 10.4 | 9.1 |
| World output ($ trillions, 1990 prices) | 43.4 | 40.7 | 78.2 | 59.3 |
| Industrial carbon emissions (GtC per year) | 8.6 | 8.4 | 9.8 | 10.0 |
| Carbon intensity (tons per $ million) | 199 | 207 | 125 | 168 |

Source for IIASA: Nakicenovic et al. 1998, appendix C. (Extrapolated from 2020 and 2050 projections.)

are very close in both 2025 and 2055. The IIASA study is a particularly useful comparison because of the detailed energy sector and great care taken in developing scenarios for different regions.

*Discussion*

Gathering the data and constructing the regional models proved the most arduous part of constructing RICE-99. The underlying economic vision is one in which nations act in a purposive manner to accumulate capital and improve future living standards. The savings rates are high in the low-income regions (with gross savings rates ranging 25 to 35 percent of GDP in low-income regions) and are between 20 and 25 percent in high-income regions. The savings rates decline in coming decades as population and economic growth decline.

The return on capital is high in developing regions, reflecting the scarcity of capital in those regions. The net return on capital in the U.S. and OECD Europe begins at the historical rate of around 5 percent per year and then declines gradually as growth slows. Rates of return in developing regions begin at 6 to 7 percent annually, then decline as their economies grow and accumulate capital.

Per capita output follows an optimistic scenario in which the four horsemen of the economic apocalypse—war, pestilence, depression, and environmental catastrophe—are largely absent. High-income regions are projected to continue growth in per capita output at around 1 percent per year over the next century. Low-income regions have per capita growth rates in the range of 2 to 3 percent annually in the next half century, with a gradual slowdown after that. Table 3.7 shows some projection but not complete convergence of developing regions toward today's living standards in the rich regions. Middle-income countries are projected to reach close to current per capita income levels in the United States by 2100. Eastern Europe and lower middle-income countries get about halfway there; low-income countries and China are projected to have per capita income levels of about ten times current levels by 2100.

The regional economic model underlying RICE-99 is one of the most complete models of the global economy available for making long-run projections and policy experiments. At the same time, many elements, particularly the assumptions for developing economies and economies in transition, are difficult to validate or estimate and are subject to large and growing projection errors as they run further into the future. It is probably impossible to provide accurate long-run projections given the rapid rate of social, economic, political, and institutional changes. Perhaps the best one can do is to heed the words of the eminent Harvard economic forecaster, Otto Eckstein, who advised that if we cannot forecast well, we should forecast often.

## Carbon Supply

RICE-99 contains a revised treatment of energy supply. In the original DICE-RICE models, the supply of fossil fuels was implicitly treated as inexhaustible. Although this was a realistic approach for the next century, it raised two major questions for the longer run. First, to the extent that fossil fuel supplies are relatively limited, this would put significant constraints on the potential total emissions and therefore on global warming over the longer run. In particular, some of the frightening scenarios put forth by William Cline[8] and others, foreseeing the potential for a 10°C long-run warming, are definitely inconsistent with standard carbon-cycle and climate models in a situation of relatively limited coal supplies. Indeed, one of the major advantages of using integrated-assessment models is that they can ensure that the different assumptions are consistent both with each other and with mainstream scientific and economic findings.

Second, and more interesting from an economic point of view, is the interaction between limited supplies and pricing. To the extent that the fossil fuel supply curve is relatively price-inelastic, prices will rise sharply as supplies are exhausted. This will lead to rising scarcity

---

8. See Cline 1992a.

prices (Hotelling rents) on carbon-based fuels. To some extent, the rising Hotelling rents would substitute for carbon taxes, and the goals for climate policies would thereby be accomplished by energy scarcity. In other words, the presence of coal-supply limits leads to the likelihood that some of the carbon tax will be shifted backwards to suppliers rather than forward to consumers. This backward shifting occurs to the extent that supply is price-inelastic. Indeed, in the limiting case of perfectly price-inelastic supply of carbon-energy with zero extraction costs, carbon taxes may have no economic effect at all and would simply redistribute rents from the resource owners to the government.

Because of this important interaction between energy supplies and climate-change policy, RICE-99 includes a simple supply relationship for fossil fuels. This sector is now described, which lumps all fossil fuels together into a single aggregate of carbon-energy. For these purposes, the one fossil fuel can usefully be thought of as coal, which is both the most abundant fossil fuel and has the predominant fraction of carbon that is potentially emitted (leaving aside the low-grade shales). We assume that carbon-energy has a limited supply with the supply curve becoming highly price-inelastic when cumulative carbon-energy use reaches 6,000 GtC (equivalent to consuming approximately 9,000 billion tons of coal). The marginal cost of carbon-energy is relatively flat until cumulative carbon-energy use is 3,000 GtC, but it quadruples (to about $450 per ton) when cumulative carbon-energy reaches 5,000 GtC. Studies by Rogner 1997 are drawn upon for both estimates of the total quantity and for the cost function for carbon-energy.

The first relationship is simply accounting for cumulative carbon-energy use:

$$CumC(t) = CumC(t-1) + 10 E(t), \tag{2.11}$$

where $CumC(t)$ is cumulative carbon-energy in GtC at the end of period $t$ and $E(t)$ is global carbon-energy use in the current period.

The important new relationship is the cost function of carbon-energy. We assume that there is a limited quantity of carbon-energy, $CumC^*$, beyond which marginal costs of extraction rise very sharply. We then assume that the marginal-cost curve takes the following form:

$$q(t) = \xi_1 + \xi_2 [CumC(t)/CumC^*]^{\xi_3}, \tag{2.12}$$

where $q(t)$ is measured in 1990 U.S. dollars per ton. Based on Rogner 1997, the numerical form of this equation in RICE-99 is:

$$q(t) = 113 + 700[CumC(t)/6000]^4. \tag{3.5}$$

The cost of carbon-energy has two terms. The first term ($\xi_1 = 113$) is the marginal cost that is independent of exhaustion. This term represents the costs of current extraction of carbon-energy today. The second term is a rising cost function. At current levels of cumulative extraction [$CumC(1995) = 0$], the second term is zero. It is highly convex, with an exponent of 4, reflecting the finding that the cost function for carbon fuels is relatively price-elastic in the near term.

The shape of the function is shown in figure 3.4. When cumulative extraction is halfway to $CumC^*$, the marginal cost of carbon-energy rises from \$113 to \$157 per ton. Carbon-energy becomes increasingly costly as the limits of resources are reached.

In some earlier versions of RICE, other constraints were added. In RICE-98, a backstop technology was added that prevented the price of carbon-energy from rising above \$500 per ton. In addition, a number of flow constraints were added in earlier versions to prevent the economy from flipping too quickly from one technology to another. These included constraints on the reduction of carbon-energy and on the introduction of the backstop constraints. In RICE-99, all these constraints were removed because they made virtually no difference to the results for the first century and added considerable computational

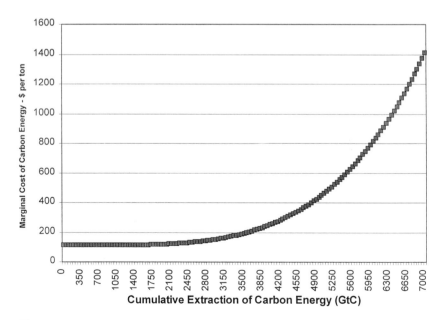

**Figure 3.4**
Carbon supply function in RICE-99 model

complexity. Users who are interested in evaluating longer-term projections should be warned, however, that because of the absence of flow constraints and a backstop technology, unreasonable outcomes and unrealistically sharp transitions can occur outside the region over which the model was calibrated and will definitely occur in the model as carbon-energy becomes exhausted.

## The Carbon Cycle and Other Radiative Forcings

An important part of the DICE-RICE-99 models is integrating the economic sectors with the physical world. Greenhouse-gas emissions affect the carbon cycle as well as other atmospheric trace gases, change the radiative balance of the atmosphere, affect climate, and then feed back to affect human societies and natural ecosystems. One of the most difficult features of developing integrated assessment models like the DICE or RICE models has been to find parsimonious relationships between economic activity and climate change. Economists are accustomed to relying on highly simplified representations of economic relationships (such as the much-used Cobb-Douglas production function), and this approach has proven fruitful in understanding phenomena ranging from business cycles to economic growth. This section describes the model used for the carbon cycle and other radiative forcings.

The use of highly simplified aggregate relationships is motivated by three practical conditions. First, an understanding of the interaction of economy and climate is advanced if the underlying structure is as simple and transparent as possible; complex systems cannot be easily understood and erratic behavior may well arise because of the interaction of complex relationships. Second, because most of the relationships in the DICE-RICE models are poorly understood, it is necessary to undertake extensive sensitivity analysis to determine the robustness of the model. The larger the model, the more difficult it is to undertake comprehensive sensitivity and uncertainty analysis. Finally, from a computational point of view, RICE-99 is already straining at the computational capacity of readily available software packages that can be used on personal computers, and the goal set here was the construction of a model that can be easily used by other researchers. In modeling, small is beautiful.

Including more sectors of the economy, more layers of the ocean, more greenhouse gases, more energy resources would reduce transparency, impair the ability to conduct sensitivity analyses, and place the model outside the envelope of current computational feasibility. To those who believe that their disciplines have been violated goes an invitation to help improve our understanding by providing better parsimonious representations of the crucial geophysical or economic processes.

This section discusses modifications of the treatment of different GHGs and of the carbon cycle in the DICE-99 model. Four major changes in the science and policy of GHGs are incorporated in RICE-99 and DICE-99.

1. In the original DICE model, $CO_2$ was aggregated with chlorofluorocarbons (CFCs) to create a $CO_2$-equivalent stock of GHGs. Since that time, the CFCs have been largely phased out in most high-income counties, and the projected radiative forcings from CFCs are consequently drastically lower than was projected in the early 1990s. Therefore, the only endogenous GHG in RICE-DICE-99 is $CO_2$.

2. The original DICE model used an empirical approach to estimating the carbon flows, relying on long-term estimates of emissions and concentrations. A number of commentators have noted that this approach may understate the long-run atmospheric retention of carbon because it assumes an infinite sink of carbon in the deep oceans. The DICE-RICE-99 models therefore replace the earlier treatment with a structural approach that uses a three-reservoir model calibrated to existing carbon-cycle models.

3. Over the last decade, climatologists have concluded that sulfates are contributing significant radiative cooling. The DICE-RICE-99 models therefore revise the treatment of exogenous anthropogenic forcing by including projections of sulfate cooling along with the positive forcings from non-$CO_2$ GHGs.

4. Projections of the radiative forcings from the non-$CO_2$ GHGs methane, $N_2O$, and tropospheric ozone have been updated over the last decade. RICE-99 forcings take into account these updates.

The next subsection discusses the structure of the carbon cycle in RICE-99. The following subsection discusses the calibration of the new carbon model, after which estimates of the radiative forcing from other GHGs are discussed.

### Revised Approach to the Carbon Cycle

In the original DICE model, the carbon cycle was estimated from time-series data on $CO_2$ emissions and concentrations. Several commentators noted that this approach may understate the long-run atmospheric retention of carbon because it assumes an infinite sink of carbon in the deep oceans. Although the original approach was reasonable for projections for the short run, it will provide misleading estimates of long-run concentrations. This point was emphasized in a contribution by Schultz and Kasting (S-K).[9] They indicate that the long-term projections in the original DICE model significantly understate atmospheric concentrations of $CO_2$.

In considering alternative approaches, it is desirable to have parsimonious representations, to have models that are structural (in the sense of reflecting solid scientific or economic underpinnings), and to rely on models whose essential findings are robust to changes in the specification. In the carbon cycle, the major trade-off involved is whether complicating the model with a more elaborate specification will produce more reliable results. Initial experiments with DICE, confirmed for the present analysis, suggest that current policy is largely unaffected by using a more elaborate specification in the base case. On the other hand, if the analyst is interested in long-run projections or if lower discount rates are used, the original DICE specification can be quite misleading. Because the alternative specification is relatively straightforward, it has been adopted for RICE-99.

The new treatment uses a structural approach with a three-reservoir model calibrated to existing carbon-cycle models. The basic idea is that the deep oceans provide a vast but limited sink for carbon in the long run. In the new specification, it is assumed that there are three reservoirs for carbon—the atmosphere, a quickly mixing reservoir in the upper oceans and the short-term biosphere, and the deep oceans. Each of the three reservoirs is assumed to be well mixed in the short run, while the mixing between the upper reservoirs and the deep oceans is assumed to be extremely slow.

We assume that $CO_2$ accumulation and transportation can be represented as a linear three-reservoir model. (The model pertains only to $CO_2$, as other GHGs are taken as exogenous.) Let:

$M_i(t)$ = total mass of carbon in reservoir $i$ at time $t$ (GtC);

$\phi_{ij}$ = the transport rate from reservoir $i$ to reservoir $j$ per unit time.

The reservoirs are $AT$ = atmosphere, $UP$ = all quickly mixing reservoirs (the upper level of the ocean down to 100 meters and the relevant parts of the biosphere), and $LO$ = deep oceans.

---

9. See Schultz and Kasting 1997.

Here are the major assumptions about these dynamics: first, the carbon cycle was in equilibrium in the year 1750; second, all emissions are into the atmosphere; third, there are no flows between the atmosphere and the deep ocean.

The dynamics of this system are as follows:

$$M_{AT}(t) = 10 \times ET(t-1) + \phi_{11}M_{AT}(t-1) + \phi_{21}M_{UP}(t-1). \tag{3.8}$$

$$M_{UP}(t) = \phi_{22}M_{UP}(t-1) + \phi_{12}M_{AT}(t-1) + \phi_{32}M_{LO}(t-1). \tag{3.9}$$

$$M_{LO}(t) = \phi_{33}M_{LO}(t-1) + \phi_{23}M_{UP}(t-1). \tag{3.10}$$

The carbon cycle has a built-in lag in which the stock of carbon in period $t + 1$ reflects the emissions in period $t$. The interpretation is therefore that the carbon stock is measured at the beginning of the period. The temperature also has a one-period lag.

## Calibration

This subsection describes the calibration of the model in equations (3.8) to (3.10). In the original DICE model, the short-run coefficients were estimated from time-series data and the long-run coefficient was derived from estimates of the adjustment time of the deep oceans.

Schultz and Kasting (1997) show that the approach in the original DICE model significantly underestimates long-run atmospheric concentrations. RICE-99 therefore modifies the original DICE model using the three-reservoir approach laid out above and calibrates it to the Bern carbon cycle model with a neutral biosphere (a neutral biosphere is one in which changing atmospheric concentrations of $CO_2$ do not change the mass of carbon in the biosphere).[10]

More precisely, the estimates are derived as follows. We assume that the reservoirs were in equilibrium in preindustrial times.[11] For calibration, we take estimates of the impulse response function from the Bern model for 20, 40, 60, 80, and 100 years. Because of nonlinearities in the response, we calibrated the model for a concentration of two times preindustrial levels. We then choose the parameters $\phi_{12}$ and $\phi_{23}$ and the effective initial mass of the upper stratum to minimize the squared deviation of the RICE-99 impulse response function from that in the Bern model. The RICE-parameterized function fits extremely well over

---

10. See IPCC 1996a, p. 86.
11. By preindustrial times, we mean the year 1750.

the period of fit of 100 years. The average absolute error in the RICE specification is 0.5 percent of the value in the Bern model.

The model then incorporates an active biosphere as follows. According to IPCC 1996a, the mass of carbon in the terrestrial biosphere in 1985 is 2190 GtC, of which 610 GtC is vegetation. We assume that only the vegetation responds to elevated atmospheric carbon, and that the elasticity of biomass in vegetation with respect to atmospheric concentrations of carbon is 0.5 (this being the so-called beta factor). We assume that the biosphere is neutral in the long run and therefore adjust the masses in the different reservoirs to ensure this constraint. Finally, we assume that half the oceans are so poorly mixed that they are unavailable for carbon absorption. After all adjustments, this implies that the effective masses in the atmospheric, upper reservoir, and lower reservoir in preindustrial times are calibrated to be 583, 705, and 19,200 GtC.

The transfer rates are 0.333 per decade from the atmosphere to the upper reservoir and 0.115 per decade from the upper reservoir to the lower reservoir. These figures imply a relatively small upper layer of the ocean with a relatively rapid transfer from that to the lower oceans. Carbon-cycle studies have found that carbon exchanges with a 800-year adjustment time between the upper reservoir and the lower reservoir, but that is usually estimated with a much larger upper ocean.[12]

The response functions to pulse inputs of carbon are shown in figure 3.5. As can be seen, the RICE-99 model with a neutral biosphere fits the Bern model very closely (these are the line and the squares at the top). The solid circles at the bottom show the RICE-99 model with the active biosphere, while the diamonds show the DICE-94 model. The figure shows how the original DICE model tends to underpredict atmospheric concentrations in the long run. The major uncertainty in the carbon cycle, however, is probably the extent to which the biosphere will continue to take up a substantial fraction of cumulative emissions.

Figure 3.6 compares the projections of $CO_2$ concentrations over the period 1990–2100 using the IPCC's IS92a emissions trajectory with the

---

12. This study uses "adjustment time" in the sense of "e-folding time" used in the physical sciences. This concept originates from the dynamics of processes that experience exponential decay. Suppose that a process evolves according to $dx(t)/dt = -\delta x(t)$. Starting in equilibrium with $x(0) = 0$, say there is a shock of $\epsilon$ to $x$ at $t = 0$, so $x(t) = \epsilon \exp(-\delta t)$. Therefore, when $T = 1/\delta$, $x$ has declined to $x(T) = \epsilon/e$. Hence the "e-folding time" is the time required for an exponential process to decay to $1/e = 0.37$ of its equilibrium value after a shock.

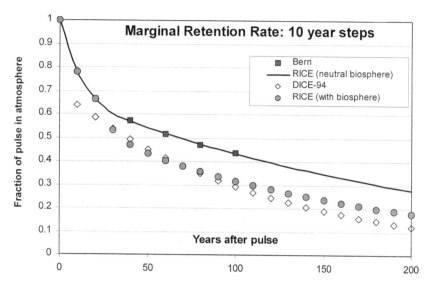

**Figure 3.5**
Impulse response functions for different models
Note: Solid line is projection from RICE-99 model assuming a neutral biosphere (that is, no effect of changing concentrations on mass of carbon in biosphere). This matches closely the simulation of the Bern model with a neutral biosphere, shown as squares. The bottom two lines show the RICE-99 and DICE-94 models with active biospheres.

Bern and RICE-99 models. RICE tends to overpredict concentrations over the next century, in part because of the simpler structure and in part because it was calibrated to the higher $CO_2$ concentrations, which implies a higher atmospheric retention in the near term.

*Other Greenhouse-Gas Emissions*

RICE-99 considers greenhouse warming primarily from carbon dioxide. Although there are large uncertainties involved, the total net radiative forcings of non-$CO_2$ GHGs and aerosols in 2100 are currently expected to be an order of magnitude smaller than those for $CO_2$. Moreover, the policy instruments available to affect gases other than the CFCs are very poorly understood at the present time. Because of their relatively small importance and the absence of clear policies to affect them, other GHGs and aerosols are assumed to be exogenous in RICE-99. Table 3.9 shows the assumptions about the radiative forcings of non-$CO_2$ GHGs.

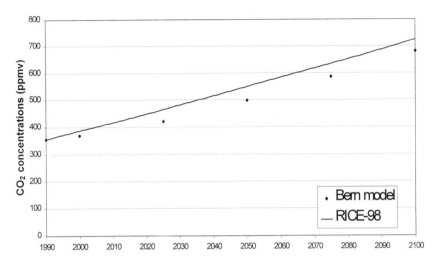

**Figure 3.6**
Comparison of projections of $CO_2$ concentrations from RICE-99 and Bern models for emissions projections
Note: Solid line is projection of $CO_2$ concentrations using IPCC IS92a emissions projection. Diamonds are projections using Bern model. Emissions and the Bern simulation are from IPCC 1996a, p. 23.

The effects of the changes in the RICE carbon cycle and projections of other GHGs and aerosols on radiative forcing in 2100 can be seen in table 1.2. The change in the carbon cycle makes little difference for a time frame as small as a century, but the reduced forcing from non-$CO_2$ GHGs has a bigger impact than the rather large reduction in baseline carbon emissions.

**The Climate Module**

Climate modelers have developed a wide variety of approaches for estimating the impact of rising GHGs on climatic variables. The models typically taken to be the most satisfactory are the large general circulation models (GCMs). These require several hundred hours of supercomputer time simply to perform a simulation, and including them in an optimization model of the kind described here is not feasible.

To develop integrated models of climate and the economy, it is necessary to have a relatively small model that links GHG concentrations and the major climatic variables. Only the impact of GHGs on global

**Table 3.9**
Non-$CO_2$ radiative forcings according to IPCC-90, MAGICC/IPCC-99, and RICE-99

A. Estimate forcings (watts per meter squared)

| Year | (1) IPCC-90 | (2) MAGICC/IPCC-99 | (3) RICE-99 |
|------|------|------|------|
| 1990 | 0.95 | | |
| 2000 | | −0.07 | −0.06 |
| 2010 | | | 0.07 |
| 2020 | | | 0.21 |
| 2030 | | | 0.34 |
| 2040 | | | 0.48 |
| 2050 | 1.85 | 0.69 | 0.61 |
| 2060 | | | 0.75 |
| 2070 | | | 0.88 |
| 2080 | | | 1.02 |
| 2090 | | | 1.15 |
| 2100 | 2.12 | 1.14 | 1.15 |

B. Components of forcings, MAGICC/IPCC-99 (watts per meter squared)

| | 2000 | 2050 | 2100 |
|------|------|------|------|
| $CO_2$ | 1.85 | 3.55 | 5.01 |
| $CH_4$ | 0.48 | 0.86 | 1.07 |
| Tropospheric ozone | 0.41 | 0.49 | 0.55 |
| Halocarbons and stratospheric ozone | 0.16 | 0.24 | 0.34 |
| $N_2O$ | 0.17 | 0.23 | 0.24 |
| Aerosols (direct and indirect) | −1.29 | −1.13 | −1.06 |
| Total | 1.78 | 4.24 | 6.15 |
| Total, non-$CO_2$ | −0.07 | 0.69 | 1.14 |

Sources: Part A, column (1) is from IPCC 1990, pages 54 and 57.
Part A, column (2) and part B are the result of the MAGICC model (Wigley, Solomon, and Raper 1994, taking emissions from IPCC.
Scenario B2 as implemented in IIASA's MESSAGE model as of Jan. 11, 1999.
Part B is from the same source as part A, column (2).

mean temperature are included in RICE-99. Although this analysis focuses primarily upon globally averaged surface temperature, it is recognized that this variable is not the most important for impacts. Variables like precipitation or water flows—along with extremes of droughts, floods, and freezes—are more important for economic activity than is average temperature alone. Mean temperature is chosen because it is a useful *index* of climate change that tends to be associated with most other important changes. In the language of statistics, temperature is likely to be a sufficient statistic for the other variables that have an important impact upon human and natural societies. This point can be seen in surveys of GCMs, in which the estimated impact of $CO_2$ doubling on mean temperature is highly correlated with the impact on precipitation.

RICE-99 takes the same approach as the one developed in the original DICE model. This uses a simplified minimodel to represent the basic dynamics of climate change. It then uses larger models to calibrate the major parameters of the minimodel. It must be emphasized that this representation is highly simplified and is intended only to depict the broad features of climate change.

The description of the climate model will be extremely abbreviated because the specification used here is identical to that used in the original DICE model. No changes in climatology or GCM models have appeared that would lead to revisions in either the specification or the parameters of the model.

The approach here follows closely the model developed by Schneider and Thompson 1981. The climate system is represented by a multistratum system, including an atmosphere, an upper-ocean stratum, and a lower-ocean stratum. The system has an atmosphere that is warmed by solar radiation and is in short-run radiative equilibrium. The accumulation of GHGs warms the atmosphere, which then mixes with and warms the upper ocean, which in turn heats the deep oceans. The atmosphere exchanges energy quickly with the upper oceans, which impose a certain amount of thermal inertia on the system because of their heat capacity. The upper stratum of the ocean also exchanges water with the lower stratum, representing the deep oceans, and the rate of heat transfer is proportional to the rate of water exchange. This model is a box-advection model, which is simpler to include in economic models than the mixed box-advection and upwelling-diffusion approach that is widely used in medium- and large-scale models today. The two state variables in the two-equation model are the globally averaged surface temperature and the deep-ocean temperature.

The equations of the model, given in chapter 2, are:

$$T(t) = T(t-1) + \sigma_1 \{F(t) - \lambda T(t-1) - \sigma_2[T(t-1) - T_{LO}(t-1)]\}. \qquad (2.15a)$$

$$T_{LO}(t) = T_{LO}(t-1) + \sigma_3[T(t-1) - T_{LO}(t-1)]. \qquad (2.15b)$$

In this model, $(1/\sigma_1)$ represents the thermal capacity of the atmospheric layer and the upper oceans, $(1/\sigma_3)$ is the transfer rate from the upper level of the ocean to the deep oceans, $\sigma_2$ is the ratio of the thermal capacity of the deep oceans to the transfer rate from shallow to deep ocean. A key parameter in all models is $\lambda$, or the feedback parameter. This parameter is a way of representing the equilibrium impact of $CO_2$ doubling on climate. By solving equation (2.12) for a constant temperature, it is easily seen that the long-run or equilibrium impact of a change in radiative forcing is $\Delta T / \Delta F = 1 / \lambda$. We use the parameter $T_{2 \times CO_2}$ to represent the equilibrium impact of doubled $CO_2$ concentrations on global mean surface temperature. From equation (2.12), therefore, we have that $T_{2 \times CO_2} = \Delta F_{2 \times CO_2} / \lambda$, where $\Delta F_{2 \times CO_2}$ is the change in radiative forcing induced by a $CO_2$ doubling. The derivation of $T_{2 \times CO_2}$ is given in numerous sources.

For calibration purposes, three different models have been examined. The first is the Schneider-Thompson (ST) approach 1981. This study develops a two-equation model that is identical to equation set (2.12); it has the disadvantage of being highly simplified relative to larger models. To exploit the ST approach, we construct the model explicitly using the parameters developed in the original study. The second, the most completely developed model examined, was a coupled atmospheric-ocean model developed by Stouffer, Manabe, and Bryan 1989. This model is a highly disaggregated representation of both the atmosphere and the oceans, and it provides a transient calculation of the impact of slowly rising $CO_2$ concentrations. A third model, much in the spirit of the approach used here, is a parametric representation of the Oregon State University model in a small model of the coupled atmospheric and six-layer ocean model developed by Schlesinger and Jiang 1990. This model uses the larger model to determine the parameters of the smaller model and then uses the smaller model for calculating transient values over longer periods.

The three models gave similar trajectories. The original DICE and RICE models used the calibration of the SJ model because that appeared closest to that used by expert groups of the U.S. National Academy of Sciences and of the IPCC. This model has an equilibrium warming of $T_{2xCO_2} = 4.1/\lambda = 2.91°C$ for a $CO_2$ doubling, with an e-fold time for temperature of thirty years. A full discussion is contained in Nordhaus 1994b. Although developments in climate modeling have not revised estimates of the impact of rising GHGs on equilibrium climate change, recent assessments have noted that climate models tend to overpredict the extent of global warming over the last two decades. For example, in the RICE-99 model, using actual concentrations of $CO_2$ in the atmosphere would produce an increase in global mean temperature for the 1990s that is almost twice the observed increase relative to the first half of the twentieth century. In light of the overpredict ion from climate models, $T_{2xCO_2}$ is often taken to be 2.5°C in policy evaluations. A model-based estimate has been chosen here because of the likelihood that many confounding factors are involved in the overprediction of warming to date.

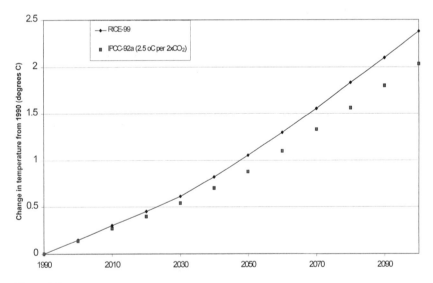

**Figure 3.7**
Comparison of temperature simulation of RICE-99 model with IPCC-96
Note: Both calculations use the radiative forcings with aerosols and emissions according to IPCC IS92a from IPCC 1996a, pp. 321. The IPCC calculation has an equilibrium temperature increase of 2.5°C for an equilibrium $CO_2$ doubling, whereas the RICE model has an equilibrium temperature increase of 2.9°C for an equilibrium $CO_2$ doubling.

Figure 3.7 compares the projections of RICE-99 with the IPCC calculations in IPCC [1996a]. Each projection uses the radiative forcing estimated according to the IS92a scenario, as updated for other non-$CO_2$ forcings in IPCC [1996a]. The RICE-99 model lies above the IPCC primarily because RICE uses a temperature-$CO_2$ sensitivity of 2.9°C, as discussed above, whereas the IPCC run has a temperature-$CO_2$ sensitivity of 2.5°C.

# 4    The Impacts of Climate Change

## Early Impact Studies

Sensible policies on global warming should weigh the costs of slowing climate change against the benefits of slower climate change. Ironically, recent policy initiatives, such as the Kyoto Protocol of 1997, have been introduced without any attempt to link the emissions controls with the benefits of the lower emissions. In part, the decoupling of policy from the benefits of the policies comes because many environmental advocates are skeptical of the use of cost-benefit analysis. Many economists are troubled by the lack of clear and convincing estimates of the impacts and by the prospect that global warming may trigger unpredictable and potentially catastrophic impacts.

This chapter provides a new set of estimates of the economic impacts of climate change. Although the literature in this area is extensive, there are many gaps in coverage of sectors and countries, and many of the most important impacts have not been satisfactorily quantified and monetized. Notwithstanding the imprecision of the estimates, it is essential that impacts be considered in the climate-change debate.

Starting with Nordhaus 1990a and 1991a, assessments of the impacts of climate change have been organized in the framework of national economic accounts with additions to reflect nonmarket activity. These first-generation impact studies for the United States were summarized in the IPCC survey and are shown in table 4.1. Most estimates find monetized damages for the United States (with its current economic structure) to lie between 1 and 1.5 percent of GDP for a 2.5 to 3°C warming. Other surveys provide estimates of the impacts for other regions and for the world (see especially the survey in IPCC 1996c).

Four points are worth noting about the first-generation studies. First, there is a deceptive degree of consensus to the estimates. They add up

**Table 4.1**
Estimated impact of climate change on the United States from 1996 IPCC report (billions of 1990 U.S. dollars)

| | Cline (2.5°C) | Fankhauser (2.5°C) | Nordhaus (3°C) | Titus (4°C) | Tol (2.5°C) |
|---|---|---|---|---|---|
| Agriculture | 17.5 | 3.4 | 1.1 | 1.2 | 10.0 |
| Forest loss | 3.3 | 0.7 | (a) | 43.6 | (a) |
| Species loss | 4.0 | 1.4 | (a) | (a) | 5.0 |
| Sea level rise | 7.0 | 9.0 | 12.2 | 5.7 | 8.5 |
| Electricity | 11.2 | 7.9 | 1.1 | 5.6 | (a) |
| Nonelectric heating | −1.3 | (a) | (a) | (a) | (a) |
| Mobile air conditioning | (a) | (a) | (a) | 2.5 | (a) |
| Human amenity | (a) | (a) | | (a) | 12.0 |
| Human mortality and morbidity | 5.8 | 11.4 | | 9.4 | 37.4 |
| Migration | 0.5 | 0.6 | | (a) | 1.0 |
| Hurricanes | 0.8 | 0.2 | 0.75% of GDP | (a) | 0.3 |
| Leisure activities | 1.7 | (a) | | (a) | (a) |
| Water supply | | | | | |
|   Availability | 7.0 | 15.6 | | 11.4 | (a) |
|   Pollution | (a) | (a) | | 32.6 | (a) |
| Urban infrastructure | 0.1 | (a) | | (a) | (a) |
| Air pollution | 3.5 | 7.3 | | 27.2 | (a) |
| Total | | | | | |
|   Billions | 61.1 | 69.5 | 55.5 | 139.2 | 74.2 |
|   Percent of GDP | 1.1 | 1.3 | 1 | 2.5 | 1.5 |

Source: IPCC 1996c.
Note: (a) are items that are not assessed or quantified or are judged to be small.

to quite similar amounts, but the details are highly divergent. Second, many of the earliest estimates (particularly those for agriculture, sea-level rise, and energy) were extremely pessimistic about the economic impacts, whereas more recent studies, which include adaptation, do not paint such a gloomy picture.[1] Third, coverage of regions outside the United States is extremely sparse. There is little serious work on other major regions of the globe, such as China, India, or Africa; this is particularly troubling because the impacts of climate change may well

---

1. See particularly the study by Darwin et al. 1995 on agriculture, by Rosenthal et al. 1994 on energy, and by Yohe and Schlesinger 1998 on sea-level rise. The recent study of impacts for the United States by Mendelsohn and Neumann 1999 contains the most optimistic outlook of recent research.

be largest in those regions. Finally, and most important, many of the most pressing concerns about global warming, particularly the concern with catastrophic risk, have not been adequately studied.

In reviewing current research, it is clear that the results are highly conjectural and that it continues to be difficult to make solid estimates of the impacts of climate change. Of all the sectors, only agriculture and sea-level change have made significant progress in estimating climate-change impacts on a detailed regional level. In some areas, such as ecosystems and human health, the difficulties are particularly formidable because of enormous uncertainties about the underlying physical and biological impacts and about the potential for adaptation. For concerns such as ecosystems, valuation is extremely difficult, while there are no established methodologies for valuing catastrophic risks.

### The Present Approach

The present study follows first-generation approaches by analyzing impacts on a sectoral basis. There are three major differences from many earlier studies. First, the approach here focuses on all regions rather than concentrating on the United States. This focus is obviously necessary both because global warming is a global problem and because of a common view that impacts are likely to be significantly larger in poorer countries. Second, this study focuses more heavily on the nonmarket aspects of climate change because of the finding of the first-generation studies that the impacts on market sectors outside of agriculture are likely to be relatively limited.

The final difference from earlier studies is that the present approach relies on a willingness to pay (WTP) approach to estimating the value of preventing future climate change. This approach is taken because the authors believe that comprehensive regional estimates of impacts are unlikely to be available in the near future. Moreover, scholars are gravitating toward the view that it is the *risks* that are the major cause for concern about future climate change. The WTP approach estimates the "insurance premium" that different societies are willing to pay to prevent climate change and its associated impacts. The advantage of the WTP approach is that it can encompass different approaches to measuring impacts, including surveys as well as statistical impact measures.

This chapter presents the results for thirteen subregions of the eight regions of RICE-99. These subregions were originally selected for a more disaggregated version of RICE used in RICE-98. Discussion of the estimates for all thirteen subregions will be particularly useful for those researchers who would like more disaggregated estimates. The breakdown of the RICE-99 regions into the subregions of the impacts analysis is shown in table 4.2. Table 4.3 shows the current mean annual temperature of the thirteen subregions. For these, both area-weighted and population-weighted climates have been provided. The differences between these two concepts are particularly dramatic for countries like the United States, Canada, and Russia.

In estimating impacts, the potential areas of concern have been divided into seven categories:

1. Agriculture

2. Sea-level rise

3. Other market sectors

4. Health

5. Nonmarket amenity impacts

**Table 4.2**
Regions in impact analysis

| Region in 8-region RICE 99 model | Subregion in 13-region impact assessment | Description |
| --- | --- | --- |
| United States | United States | United States |
| Other high income | Japan | Japan |
| | Other high income | RICE-99 "OHI" without Japan |
| OECD Europe | OECD Europe | OECD Europe |
| Russia and Eastern Europe | Eastern Europe | RICE-99 "Russia and Eastern Europe" without Russia |
| | Russia | Russia |
| Middle income | Middle income | RICE-99 "MI" without high-income OPEC |
| | High-income OPEC | OPEC members with per capita incomes over $6,000 |
| Lower middle income | Lower middle income | RICE-99 "LMI" |
| China | China | China |
| Low income | India | India |
| | Africa | Sub-Saharan low-income countries |
| | Low income | RICE-99 LI without India and Africa |

**Table 4.3**
Subregional mean temperature (annual average, degree C)

| Subregion | Area-weighted | Population-weighted |
|---|---|---|
| Russia | −5.0 | 3.1 |
| Eastern Europe | 7.9 | 8.9 |
| OECD Europe | 9.1 | 10.5 |
| Other high income | 6.6 | 12.5 |
| Japan | 8.9 | 12.6 |
| United States | 7.8 | 13.4 |
| China | 4.6 | 13.8 |
| Lower middle income | 19.2 | 19.0 |
| Middle income | 21.7 | 21.3 |
| Low income | 21.2 | 22.6 |
| High-income OPEC | 23.3 | 24.0 |
| Africa | 24.3 | 24.2 |
| India | 22.4 | 25.7 |

Source: Prepared by the authors. Both columns are averaged over 1 degree by 1 degree gird.

6. Human settlements and ecosystems

7. Catastrophes

The methodology for estimating impacts is as follows. Let $\theta_{ij}(T, y_j)$ represent the impact index for sector $i$ in subregion $j$ for a global temperature increase of $T$ and a per capita income of $y_j$. This is the fraction of annual output that subregion $j$ would be willing to pay to avoid the consequences on sector $i$ of a temperature increase of $T°$C. It is a function of $y_j$, subregional output per capita. To determine the impact of warming in a particular future year, we assume that the future impact index takes the following form:

$$\theta_{ij}(T, y_j) = Q_{ij}(T) \times [y_j(t)/y_j(1995)]^{\eta_i}. \tag{4.1}$$

In this formula, the impact is related to a function of temperature times an income adjustment. The adjustment is the ratio of per capita GDP in the future year to present per capita GDP $[y_j(t)/y_j(1995)]$ to the power $\eta_i$, where $\eta_i$ is the income elasticity of the impact index.

United States agriculture can serve here as an example. Our estimate is that $\theta_{agriculture,US}[2.5, y_{US}(1995)]$ is 0.065 percent. This means that the absolute dollar value of the negative impact on agriculture of 2.5°C global warming is estimated to be about \$4 billion in today's economy. The income elasticity of the impact index is estimated to be −0.1, based on the declining share of agriculture in output as per capita output rises. Under this assumption, if U.S. GDP per capita income doubles, the impact index will decline from 0.065 percent to 0.061 percent of GDP.

The first step in constructing the impact index $\theta_{ij}(T, y_j)$ for each category and subregion is to use available information to estimate $Q_{ij}(2.5)$. According to the base run of RICE-99, a 2.5°C rise will occur around 2100. The next step is to make an estimate of the value of $\eta$. These two steps are carried out in the third section. In the fourth section, we create the functions that allow us to evaluate $\theta_{ij}(T, y_j)$ at other temperatures. The fifth section explains how the impact indices are used to calibrate the damage functions in RICE-99 (see equations 2.16 and 2.17).

## Discussion of Individual Sectors

### Agriculture

The sector with the best underlying data and most extensive research is agriculture. The range of estimates of impacts for a benchmark 2.5°C warming, including the $CO_2$ fertilization effect, is from slightly positive to severe negative impacts. Early studies concentrated on the "production-function approach" and often found severe negative impacts, while more recent approaches incorporating extensive adaptation or using the economically oriented "Ricardian" approach find lower levels of damages. Estimates in the IPCC review (IPCC 1996c, p. 203) for the United States found a range of impacts from \$1.1 to \$17.5 billion. For other countries, Fankhauser 1995 estimated welfare losses in agriculture ranging from 0.16 percent of GDP in the United States to 2.1 percent of GDP in China for a 2.5°C warming (see IPCC 1996c, p. 204). A review by Schimmelpfennig 1996 usefully summarizes recent research.

Here, we have combined the regional estimates prepared by Darwin et al. 1995 of the share of agricultural revenues lost due to an equilibrium $CO_2$ doubling with estimates of the share of agricultural output in GDP to construct the impact index for agriculture. Darwin et al. is one of the most detailed studies, including several regions and incorporating both detailed agronomic data as well as equilibrium modeling of land use and of economic impacts. It presents results for an equilibrium $CO_2$ doubling in four GCMs (with the impact on global mean temperature ranging from a 2.8 to a 5.2°C). It excludes $CO_2$ fertilization effects as too speculative and is therefore likely to present an overestimate of damages or an underestimate of benefits. The results of the second most unfavorable GCM are used with the aim of including some loss-aversion in the WTP estimates.

The Darwin et al. study does not use the same regional breakdown as the estimates presented here. Estimates were available for the United States, Japan, Europe, and China, but other subregions had to be adjusted. For India and the middle-income subregion, we have used recent studies employing the Ricardian technique.[2] There are currently no detailed economic studies available for OPEC, lower middle-income, low-income, or African countries, and we have used the results from Darwin et al. for the "rest of world" for these subregions; similar numbers have been found in other agricultural studies for low-income regions, but there are major uncertainties here.

The assumption here is that the elasticity term in this sector is –0.1. This implies that a doubling of per capita output will reduce the share of the agricultural sector and therefore the impacts by about 7 percent. The overall estimates for agriculture are shown in table 4.4.

*Sea-Level Rise*

The coastal sector is one of the most accessible and best documented areas of impact. Estimates of sea-level rise are determined by global warming and have little regional component, so the lack of regional precision in climate models does not arise here. The studies reviewed by the IPCC (IPCC 1996c, p. 203) estimated the annualized costs of sea-level rise to the United States to be between $5.7 and $12.2 billion, this being between 0.1 and 0.2 percent of GDP. Recent studies of developed properties in the United States find relatively small impacts of sea-level rise over the next century. The first-generation studies shown in table 4.1 appear to have overestimated impacts because they were based on sea-level rise estimates that were much larger than current projections.

Second-generation studies find relatively modest estimates of pure sea-level rise for the United States. Yohe and Schlesinger 1998 find an estimated present value of impacts for the United States in their base case of $0.95 billion in the expected value with foresight and $3.72

---

2. See documentation in table 4.4.

**Table 4.4**
Estimated damages on agriculture from $CO_2$ doubling (benefits are negative while damages are positive)

|  | Billions, 1990 U.S. dollars | % of GDP |
|---|---|---|
| United States (a) | 3.90 | 0.07 |
| China (a, b) | –3.00 | –0.51 |
| Japan (a) | –17.20 | –0.55 |
| OECD Europe (a) | 42.10 | 0.58 |
| Russia (c) | –2.88 | –0.87 |
| India (d) | 5.11 | 1.54 |
| Other high income (a, e) | –10.40 | –1.14 |
| High-income OPEC (f) | 0.00 | 0.00 |
| Eastern Europe (g) | 2.26 | 0.58 |
| Middle income (h) | 19.51 | 1.43 |
| Lower middle income (i) | 0.65 | 0.06 |
| Africa (i) | 0.10 | 0.06 |
| Low income (i) | 0.30 | 0.06 |

Notes: (a) Darwin et al. 1995, table B6. Uses second most unfavorable outcome for unrestricted case on land use.
(b) "Other East Asia" in Darwin et al. study.
(c) Uses estimate for Canada, but is similar to finding in Kane et al. as cited in Schimmelpfenning 1996, p. 31. Estimated impact on Canada is benefit of $11.3 billion on an agricultural production of $91.1 billion. Share of agriculture in Russian GDP is 7%, so impact is –(11.3/91.1)(0.07)(GDP).
(d) Uses Ricardian estimate from Dinar et al. 1998. Uses scenario III and pooled cross section from table 5.5.
(e) Canada, Australia, and New Zealand from Darwin et al. and authors' assumptions.
(f) Authors' assumptions.
(g) Assumed same percentage of GDP as OECD Europe.
(h) Uses estimates for Brazil from Sanghi et al. 1997. Assumed loss is 13% of agricultural output.
(i) No detailed estimates were available. Uses Darwin et al. estimate for "Rest of World."

billion in the ninetieth percentile case without foresight.[3] Annualizing this at an annuity factor of 3 percent per year, this yields a cost of 0.0005 to 0.0019 percent of income. Estimates of the transient costs from figure 7 in Yohe and Schlesinger divided by our estimated GDP in 2100 are virtually identical.

These estimates exclude three factors: storms, undeveloped land, and the cost of resettlements. Current estimates of these are tenuous. The storm component is most easily understood. During the period

3. See Yohe and Schlesinger 1998, table 4.

1987–95, the damage from major tropical storms in the United States averaged 0.083 percent of GDP.[4] If storm damage were to double, this would be a noticeable impact. The impact on undeveloped land is likely to be smaller than that for developed land both because natural adaptation can more easily take place since the land value is lower and there are minimal vulnerable structures. The settlements component is included in "human settlements and ecosystems."

Weighing all these factors, we assume that an estimate of 0.1 percent of income is a reasonable WTP estimate for preventing a 2.5°C warming for the coastal sector of the United States. This would have a major storm component and would assume that the upper limit of costs for developed and undeveloped land were used.

There is surprisingly little work on the impact of sea level rise outside the United States. Therefore, our estimates have been adapted using an index of coastal activity that takes into account the coastline of different subregions. We begin with an index of coastal vulnerability, which is the coastal area to total land area ratio divided by the same ratio for the United States. The coastal area is defined as the area of a subregion within 10 kilometers of the coastline. Each subregion's $Q_{coastal,j}(2.5)$ is then equal to the U.S. vulnerability times the subregion's coastal vulnerability index. It is assumed that this value has an income elasticity of 0.2 to reflect the rising urbanization and rising land values with higher per capita incomes. These estimates are shown in table 4.5.

*Other Vulnerable Market Sectors*

Other market sectors have been found relatively unaffected by current climate differences and therefore are estimated to be relatively invulnerable to climate change (ignoring indirect impacts through other sectors). A few sectors are moderately vulnerable to climate change, including forestry, energy systems, water systems, construction, fisheries, and outdoor recreation. Table 4.6 shows a breakdown of the U.S. economy by sectors of vulnerability in 1994. This indicates that vulnerable sectors comprise about 6 percent of the economy: of this, 1.7 percent is farms and related; 0.9 percent is coastal related; and the balance of 3.3 percent is distributed among other sectors. More than 90 percent of the economy, however, is not likely to be significantly affected by climate change.

---

4. Statistical Abstract 1997, p. 240.

**Table 4.5**
Coastal vulnerability

|                        | Coastal index (a) | Coastal impact (% of GDP, 1990) |
| ---------------------- | ----------------- | ------------------------------- |
| United States          | 1.00              | 0.10                            |
| China                  | 0.71              | 0.07                            |
| Japan                  | 4.69              | 0.47                            |
| Western Europe         | 5.16              | 0.52                            |
| Russia                 | 0.94              | 0.09                            |
| India                  | 1.00              | 0.10                            |
| Other high income      | 1.41              | 0.14                            |
| High-income OPEC       | 0.52              | 0.05                            |
| Eastern Europe         | 0.14              | 0.01                            |
| Middle income          | 0.41              | 0.04                            |
| Lower middle income    | 0.94              | 0.09                            |
| Africa                 | 0.23              | 0.02                            |
| Low income             | 0.94              | 0.09                            |

Note: (a) Ratio of fraction of area in coastal zone in country to that fraction in the United States. Coastal zone is defined as that part of the region that lies within 10 kilometers of an ocean.

Earlier studies contain little consistent evidence on these sectors. For the United States, Cline 1992a found losses about 0.2 percent of GDP, Nordhaus 1991a estimated impacts of essentially zero, while Mendelsohn and Neumann 1999 found small benefits in these sectors. The two sectors that have been carefully studied—forestry and energy use—show close to zero loss for the United States for a benchmark 2.5°C warming.

In light of the lack of impact turned up by current studies, the impact in this sector is rated at zero in temperate climates. For cold climates, it is estimated that energy expenditures will decline by 5 percent to reflect declining demand for heating, while in tropical and semitropical subregions it is estimated the demand for cooling will increase energy expenditures by 8 percent for incomes at current U.S. levels. Reflecting the high income elasticity of expenditures on cooling, the overall income elasticity of the vulnerability is estimated to be 0.2.

## Health

Potential impacts on human health are one of the major concerns about global warming. Studies indicate the potential for the spread of

**Table 4.6**
Vulnerability of economy to climate change (U.S. economy as of 1994)

| Sector | Gross domestic product (billions) | Percent of total | |
|---|---|---|---|
| *Gross domestic product* | 6,931.4 | 100.0 | 100.0 |
| *Major potential impact* | | | 1.7 |
| Farms | 82.2 | 1.2 | |
| Agricultural services, forestry, fisheries | 35.7 | 0.5 | |
| *Moderate potential impact* | | | 4.2 |
| Water transportation | 10.6 | 0.2 | |
| Real estate: coastal property (a) | 60.5 | 0.9 | |
| Hotels and other lodging places | 56.1 | 0.8 | |
| Outdoor recreation (b) | 81.2 | 1.2 | |
| Energy (c) | 82.3 | 1.2 | |
| *Negligible impact* | | | 94.1 |
| Mining | 90.1 | 1.3 | |
| Construction | 269.2 | 3.9 | |
| Manufacturing | 1,197.1 | 17.3 | |
| Transportation, communication, and public utilities less moderate impact | 513.5 | 7.4 | |
| Finance, insurance, and real estate less moderate impact | 1,213.2 | 17.5 | |
| Trade, wholesale and retail | 1,071.8 | 15.5 | |
| Services less hotels and recreation | 1,205.4 | 17.4 | |
| Government and statistical discrepancy | 962.6 | 13.9 | |

Sources: Data are based on the U.S. National Income and Product Accounts, Survey of Current Business, August 1996.
Notes: "Major potential impact" is more than 10 percent of output from 2.5°C global warming. "Moderate potential impact" is between 2 and 10 percent of output from 2.5° C global warming. "Negligible impact" is impact less than 2 percent of output from 2.5° C global warming.
(a) Assumes that 10 percent of real estate is vulnerable to sea-level rise and other impacts of climate change.
(b) Hunting, fishing, boating, golf, and national park expenditures, less lodging.
(c) Estimate of energy use for heating and cooling and hydroelectric power.

tropical diseases to subtropical or temperate regions if warming proceeds more rapidly than improvements in health care can keep pace. Among the major tropical vector-borne diseases that may increase their range are malaria, dengue, and yellow fever.[5] Impacts may also occur through the interaction of air and water pollution with higher temperature and more frequent river floods.

There are currently no comprehensive studies of the health impacts of global warming. Existing studies have examined heat stress in the United States associated with heat waves, but these are basically irrelevant in determining the impact of global warming because they examine short-term impacts (i.e., deviations from base climates) rather than the trend (i.e., changes in climates). One anomaly in current studies is that they find an adverse impact of heat waves on mortality and morbidity in the United States, even though the cross-sectional association between average temperature and mortality from heat stress in the United States is negative rather than positive. Most studies also ignore the impacts of extremes of cold (which would be reduced) and focus primarily on extremes of heat (which would be increased).

In the absence of systematic estimates of health impacts, we have relied on estimates based on the current prevalence of climate-related diseases. The most comprehensive study of the global incidence of disease provides estimates of years of life lost (YLLs) and disability adjusted lives lost (DALYs) prepared by Murray and Lopez 1996. Based on the data in that study, we have classified diseases into climate related and nonclimate related. The former include malaria, along with a broad group of tropical diseases, dengue fever, and pollution. Murray and Lopez group subregions into eight broad regions that correspond reasonably well with the grouping in RICE. Table 4.7 shows the YLLs from different climate-related diseases. Among regions, the climate-related years of life lost are quite small (0.63 percent of YLLs) in established market economies, but they rise to a significant fraction (11.76 percent of YLLs) in Sub-Saharan Africa. Murray and Lopez also estimate baseline improvements in each disease and region over the 1990–2020 period.

The impact of climate change has been estimated here using three approaches. Method A assumes that one-half of the decreases in YLLs for climate-related diseases estimated by Murray and Lopez over the 1990–2020 period will be lost as a result of a 2.5° C warming. Method

5. IPCC 1996b, p. 572.

**Table 4.7**
Years life lost (YLL) from climate-related diseases

| Disease | Years life lost (1000's) | | | | | | | | | |
|---|---|---|---|---|---|---|---|---|---|---|
| | World | EME | FSE | India | China | OAI | SSA | LAC | MEC |
| All | 906,501 | 49,674 | 35,930 | 200,059 | 117,802 | 114,592 | 226,890 | 56,240 | 105,234 |
| Climate-related | | | | | | | | | |
| Malaria | 28,038 | 2 | 0 | 769 | 8 | 2,277 | 24,385 | 392 | 204 |
| Tropical cluster | 3,430 | 1 | 0 | 1,124 | 17 | 6 | 1,781 | 332 | 109 |
| Dengue | 743 | 0 | 0 | 440 | 29 | 252 | 21 | 1 | 0 |
| Pollution | 5,625 | 310 | 1,320 | 1,267 | 549 | 600 | 490 | 377 | 711 |
| YLL from climate-related | 37,836 | 313 | 1,320 | 3,600 | 603 | 3,315 | 26,677 | 1,102 | 1,024 |
| Percent from climate-related | 4.17 | 0.63 | 3.67 | 1.80 | 0.51 | 2.74 | 11.76 | 1.96 | 0.97 |
| Regional mean temperature (degree C) | | 12.1 | 7.0 | 25.7 | 13.8 | 11.1 | 25.0 | 20.1 | 18.9 |

Sources: Disease data: Murray and Lopez (1996). Regional mean temperature: prepared by authors, population-weighted average over 1 degree by 1 degree grid.
Notes:
Regional abbreviations:
EME: established market economies
FSE: formerly socialist economies of Europe
OAI: Other Asia and Islands
SSA: Sub-Saharan Africa
LAC: Latin America and Caribbean
MEC: Middle Eastern Crescent
"Tropical cluster" includes trypanosomiasis, Chagas disease, schistosomiasis, leishmaniasis, lymphatic filariasis, and onchocerciasis.

B judgmentally adjusts the change in YLLs for each subregion to approximate the difference among subregions that is climate related. The final method C, which we will call the regression method, calculates the impact of warming on YLLs using the coefficients from a regression of the logarithm of climate-related YLLs on mean regional temperature estimated from the data presented in Murray and Lopez.[6] To value YLLs, we assume that a YLL is worth two years of per capita income.[7]

The results, shown in table 4.8, depict considerable vulnerability in Sub-Saharan Africa and somewhat lower vulnerability in India and other low-income countries. The impact on high-income subregions comes largely through pollution rather than tropical diseases. Except for Africa, the average of the three methods have been used for estimating health impacts in RICE-99. The African estimates have been adjusted downward from 4.6 to 3.0 percent to reflect the likelihood that the regression approach overestimates the impact for Africa because it assumes that the impact will be proportional to the current incidence of climatically related diseases, which is expected to fall with improvements in public health in that subregion. The income elasticity is assumed to be zero in these calculations. The impact for the thirteen subregions is determined by assigning each subregion to the Murray-Lopez region with which it has the most overlap.

### Nonmarket Amenity Impacts

Nonmarket sectors (other than health) include a wide variety of potential sources of warming damages. There is very little empirical evidence on the importance of nonmarket sectors, even for the United States. The major economic component of (human) nonmarket activities is time use. Comprehensive measures of national income have estimated that nonmarket time use has a total value close to that of all market activity.[8] It is useful, therefore, to concentrate on amenity value of climate and nonmarket time use in examining the impact of climate change. Other key nonmarket sectors are dealt with in the sections on health and human settlements and ecosystems.

---

6. The equation is the health impacts divided by GDP as a function of regional temperature.

7. See Tolley et al. 1994 for a discussion of valuation of life-years.

8. See Nordhaus and Tobin 1972 and Eisner 1989 for illustrative calculations.

**Table 4.8**
Impact of global warming on climate-related diseases (value of years of life lost due to 2.5 deg C warming as percent of GDP)

| Estimation technique | World | EME | FSE | India | China | OAI | SSA | LAC | MEC |
|---|---|---|---|---|---|---|---|---|---|
| A | 0.57 | 0.01 | 0.08 | 0.54 | 0.04 | 0.45 | 2.52 | 0.25 | 0.14 |
| B | 0.72 | 0.04 | 0.11 | 0.73 | 0.19 | 0.84 | 2.53 | 0.33 | 0.34 |
| C | 1.14 | 0.01 | 0.08 | 0.79 | 0.05 | 0.69 | 8.78 | 0.40 | 0.20 |
| Average | 0.81 | 0.02 | 0.09 | 0.69 | 0.09 | 0.66 | 4.61 | 0.33 | 0.23 |
| Coefficient of variation | 0.30 | 0.75 | 0.16 | 0.16 | 0.75 | 0.24 | 0.64 | 0.18 | 0.37 |

Source: See text.
Notes:
Regions are defined in table 4.7.
*Method A*: Assumes that one-half of projected gains in health status over 1990–2020 period are lost because of 2.5 deg C warming
*Method B*: Adjusts the climate-related health impacts to reflect different regional vulnerabilities to climate change in health sector.
*Method C*: Estimates the impact of temperature on the logarithm of climate-related illnesses. Estimated impact is the effect of 2.5 deg C warming.
Note that all techniques value years of life lost at a price of two years of income per year of life lost.

Leisure activities are among the important potential costs tabulated in the IPCC report (IPCC 1996c). In addition, amenity values of climate have been estimated for a number of countries. One approach to estimating these values would be to examine the climate-sensitivity of nonmarket time. Nordhaus tabulated nonmarket time use from surveys of time use by the University of Michigan in 1975 and 1981 for the United States on whether they were climate sensitive or nonclimate sensitive. This tabulation found that the share of climate-sensitive time use was less than 5 percent of nonmarket time. Robinson and Godbey 1997 have surveyed time use by Americans. They report that of 39.4 hours of free time per week in 1985, approximately 2.2 hours are spent in what might be climate-sensitive activities (recreation, sports, outdoors). Activities that can be definitely identified as outdoor recreation (outdoor recreation and walking and hiking) total 0.77 hours per week, or about 2 percent of free time.[9] Of the outdoor activities, most are generally enhanced by a warmer climate. A recent survey of those participating in selected sports activities found that of the estimated 235 million participants, only 24.3 million participated in activities (skiing and hockey) that would presumptively be adversely affected by warming.[10] The small fraction of time that is climate-sensitive for the United States tends to support the view that amenity values of climate are relatively small and may be positive.

A recent study by Nordhaus examines the value of climate-related time use in the United States.[11] This study relies on extremely detailed individual diaries on time use for approximately 100 different activities (such as skiing, golfing, and swimming) for 1981. The study examines time use in different activities for different U.S. states and different months. Two tentative conclusions of this study are: (1) that the time spent on climate-related activities increases with warm weather (e.g., the time gains to camping outweigh the time losses to skiing), and (2) the estimated value of a 2.5°C warming on the value of time use is modest but positive for the United States. The preferred estimates are that a 2.5°C warming will have a positive amenity impact on the United States of 0.30 percent of GDP. The impact is maximized at a mean annual temperature of about 20°C, after which the amenity impacts begin to turn negative.

---

9. Robinson and Godbey 1997, appendix B.
10. Statistical Abstract 1997.
11. See Nordhaus 1998c.

There are at present no reliable and empirically based amenity estimates for other countries. For this book, we extend the estimates of nonmarket time use for the U.S. This procedure is probably more defensible for outdoor time use than for many other sectors because the reactions of humans to outdoor temperatures are largely physiological and are unlikely to be affected by technological differences. To calculate the estimates for different subregions, the quadratic temperature relationship from the U.S. study has been applied using the subregional average temperatures in table 4.3. To evaluate these impacts, we assume that outdoor hours have a value equal to average hourly earnings, that earnings are equal to the share of wages in GDP times per capita GDP, and that there are 1,500 hours per year per worker. In this sector, we assume an income elasticity of 0.

The conclusion here is that the net amenity value of climate change is likely to be positive for temperate and cold subregions and negative for warm subregions. The approach followed here requires further study and validation, particularly for tropical and semitropical subregions.

### Human Settlements and Ecosystems

One of the most important and difficult issues to evaluate is the potential for climate change to have damaging effects on human settlements and natural ecosystems. This set of issues reflects a wide variety of factors that are difficult to model and quantify but may be of major concern. The effect on human settlements includes the difficulties that arise because immobile population, cities, or cultural treasures cannot emigrate with climate change. One example is low-lying cities like Venice, which will be hard-pressed to keep up with extensive sea-level rise. Another would be low-lying countries, like Bangladesh, the Maldives, or the Netherlands, for which coping would involve major social as well as economic costs. For small countries, the need to either adapt or migrate will impose significant costs on the populations.

Even more difficult issues arise with respect to natural ecosystems. A major concern arises with respect to irreversible effects and with immobile ecosystems. Among the irreversible effects are the potential for loss of species or the destruction or deterioration of complex ecosystems. The review in IPCC 1996b indicates a wide range of vulnerability, including potential for major desertification, impacts on the cryosphere, and degradation of coastal ecology, but quantification of such impacts has not been undertaken. At present, economic valuations of these are rather wild. As IPCC 1996c states, "Existing figures are all rather speculative. There is a serious need for conceptual quantitative work in this area."[12] Contingent valuation surveys of species often obtain large impacts, but the experiments are generally poorly defined to the respondents.[13]

Given the lack of any comprehensive estimates, the authors have made rough estimates here of the extent to which the economy and other institutions are vulnerable to climate change. For different subregions, it is assumed that the capital value of climate-sensitive human settlements and natural ecosystems range from 5 to 25 percent of regional output; for the United States, this number is estimated to be 10 percent of national output, or about $500 billion in 1990. This number is estimated to be higher in Europe, in island countries like Japan, in small countries, and in countries with sensitive ecosystems. We assume that each subregion has an annual willingness to pay 1 percent of the capital value of the vulnerable system (which is one-fifth of the annualized value at a discount rate on goods of 5 percent per year) to prevent climate disruption associated with a 2.5° C rise in mean temperature.

This approach can be illustrated for the United States. Assume that one-half the value of settlements pertains to the National Parks. This would represent a capital value of $250 billion. Parks will potentially be disrupted by a climate change of 2.5 degrees C, with damage to ecosystems, wildlife, beaches, and recreational value. We estimate that the United States would be willing to pay $2.5 billion per year to prevent these effects. Similarly, Europeans might value the unique character of low-lying regions such as Venice at $100 billion. This approach implies a willingness to pay of $1 billion per year for preventing climate change (or for preventing the impacts of climate change, which for Venice are primarily water intrusion). It must be emphasized that this methodology at this stage is speculative and requires a detailed inventory and valuation of climatically sensitive regions for validation. It appears likely that the impact on climate-sensitive ecosystems and human settlements, however, is one of the most important considerations in climate-change policy.

On the basis of these assumptions, we have then made estimates of the value of settlements for each subregion that depend upon the size,

---

12. IPCC 1996c, p. 200.
13. See the discussion in IPCC 1996c, chapter 6.

mobility, and robustness of the underlying social and natural systems. There is no evidence on the impact of higher incomes on this impact. It is likely that the concern with settlements and ecosystems rises with incomes while the costs of adaptation declines with higher incomes. The higher concern about settlements by high-income countries, particularly those involved with natural ecosystems, suggests that settlements impact has a positive income elasticity. We have estimated the elasticity of the impact with respect to income to be 0.1. The estimates for each subregion are shown in table 4.10.

### Catastrophic Impacts

There are many concerns about catastrophic impacts of climate change. Among the potential severe events are a sharp rise in sea level, shifting monsoons, a runaway greenhouse effect, collapse of the West Antarctic Ice Sheet, and changing ocean currents that would have a major cooling effect on some subregions, such as OECD Europe.

To judge the importance of catastrophic impacts of climate change, a survey of experts pose the following questions:

Some people are concerned about a low-probability, high-consequence output of climate change. Assume by "high-consequence" we mean a 25 percent loss in global income indefinitely, which is approximately the loss in output during the Great Depression. (a) What is the probability of such a high-consequence outcome for scenario A, i. e., if the warming is 3 degrees C in 2090 as described above? (b) What is the probability of such a high-consequence outcome for scenario B, i. e., if the warming is 6 degrees C in 2175 as described above? (c) What is the probability of such a high-consequence outcome for scenario C, i. e., if the warming is 6 degrees C in 2090 as described above?[14]

The respondents showed greater relative concern about the large-temperature-increase and rapid-temperature-increase scenarios. The mean (median) probability of extremely unfavorable impacts was 0.6 (0.5) percent for the 3-degrees-C-in-a-century scenario A and 3.4 (2.0) percent for scenario B. The assessment of the catastrophic scenarios varied greatly across respondents and particularly across disciplines.

Developments since the survey above have heightened concerns about the risks associated with major geophysical changes, particularly those associated with potential changes in thermohaline circulation. For example, Broecker writes:

---

14. Nordhaus 1994a.

One of the major elements of today's ocean system is a conveyor-like circulation that delivers an enormous amount of tropical heat to the northern Atlantic. The record . . . indicates that this current has not run steadily, but jumped from one mode of operation to another. The changes in climate associated with these jumps have now been shown to be large, abrupt, and global. Although the exact linkages that promote such climatic changes have yet to be discovered, a case can be made that their roots must lie in the ocean's large-scale thermohaline circulation.[15]

Climatic research has uncovered many flickering events, in which wholesale reorganizations of North Atlantic and even global climate systems occurred in periods as short as a decade (Broecker 1997, Dansgaard et al. 1993, Taylor et al. 1993).

Concerns about such catastrophic events have grown with model runs that produced changes in the thermohaline circulation. Stocker and Schmittner 1997 present a number of scenarios for global temperature increases. In a rapid temperature-increase case, where global temperature rises 4°C in a century, thermohaline circulation is shut down and remains so for at least 1,000 years. A slower temperature increase, with an increase of 4°C in two centuries, leads to a weakening but eventual recovery of thermohaline circulation. Although much further work needs to be done in this area, it does suggest that the risk of major impacts rises sharply as temperature increases beyond the 2 to 3°C range.

To reflect these growing concerns, we assume the probability of a catastrophe with a 2.5°C warming is double the estimated probability for a 3°C warming from the survey, that the probability associated with a 6°C warming is double the survey estimate, and that the percentage of global income lost in a catastrophe is 20 percent higher than the figure quoted in the survey. This implies that the probability of a catastrophic impact is 1.2 percent with a 2.5°C warming and 6.8 percent with a 6°C warming. The percentage of income lost in a catastrophe is assumed to vary by subregion. We have assumed that certain subregions (such as the Indian continent and OECD Europe) are relatively more vulnerable than other subregions. For OECD Europe, which is assumed to have high vulnerability, the percentage of income lost is assumed to be double the percentage lost in the United States.

To calculate the WTP to avoid catastrophic risk, we assume that countries are averse to catastrophic risk. For this calculation, a rate of relative risk aversion of 4 is assumed. This implies that the equivalent

---

15. Broecker 1997, p. 1582.

income loss from a 1 percent probability of a 20 percent loss is 0.32 percent of income and for a 1 percent probability of a 40 percent loss is 1.21 percent of income. The range of estimates of WTP lies between 0.45 and 1.9 percent of income for a 2.5°C warming and between 2.5 and 10.8 percent of income for a 6°C warming. It is assumed that this WTP has an income elasticity of 0.1 to reflect greater aversion to catastrophic risk as incomes increase. Table 4.9 shows the derivation of the WTP for reduction of catastrophic risk.

*Summary Estimates*

The damage estimates from the different sectors collected in table 4.10 represent the impact for a 2.5°C temperature increase with 2100 incomes; in the case of catastrophic risks, the impact of a 6°C increase is also shown. Note that these numbers differ slightly from those in the earlier tables because these impacts apply to projected incomes in 2100, while the earlier tables generally refer to 1995 incomes and economic structures.

It is useful to compare these estimates with the result of the survey of experts mentioned above. The median estimate from the survey was that about one-half of the total economic impact, or 0.9 percent of income, was due to nonmarket impacts. Discussions with experts pointed largely to two impacts, health and ecosystems. No major nonmarket impacts outside these areas were identified either by the experts or in the 1996 IPCC report. The numbers derived here are close to those for nonmarket impacts in the survey.

## Impact Indices as Functions of Temperature

The section above examined estimates of climatic damages for a benchmark 2.5°C warming. The section discusses how the impact indexes are extended to other points in the temperature domain. The process necessarily extrapolates beyond existing studies because only a sparse set of estimates exists of the shape of the damage function.

**Agriculture.** Agricultural estimates are probably the most carefully studied of all estimates. Figure 4.1 shows a linear relationship between the initial subregional temperature (area weights) and the impact index, estimated from the data in tables 4.3 and 4.4. Integrating the marginal damage relationship gives a quadratic relationship between agriculture

**Table 4.9**
Willingness to pay to eliminate risk of catastrophic climate change

| | (1) | (2) | (3) | (4) | (5) | (6) |
|---|---|---|---|---|---|---|
| | Probability of catastrophic event | | Relative vulnerability | Expected loss if catastrophic event (% of GDP) | Willingness to pay to avoid catastrophic risk (% of GDP) | |
| | 2.5 deg | 6 deg | | | 2.5 deg C | 6 deg C |
| United States | 0.012 | 0.068 | 1.0 | 22.1 | 0.45 | 2.53 |
| China | 0.012 | 0.068 | 1.0 | 22.1 | 0.45 | 2.53 |
| Japan | 0.012 | 0.068 | 1.0 | 22.1 | 0.45 | 2.53 |
| OECD Europe | 0.012 | 0.068 | 2.0 | 44.2 | 1.90 | 10.79 |
| Russia | 0.012 | 0.068 | 1.5 | 33.2 | 0.94 | 5.33 |
| India | 0.012 | 0.068 | 2.0 | 44.2 | 1.90 | 10.79 |
| Other high income | 0.012 | 0.068 | 1.5 | 33.2 | 0.94 | 5.33 |
| High-income OPEC | 0.012 | 0.068 | 1.0 | 22.1 | 0.45 | 2.53 |
| Eastern Europe | 0.012 | 0.068 | 1.0 | 22.1 | 0.45 | 2.53 |
| Middle income | 0.012 | 0.068 | 1.0 | 22.1 | 0.45 | 2.53 |
| Lower middle income | 0.012 | 0.068 | 1.5 | 33.2 | 0.94 | 5.33 |
| Africa | 0.012 | 0.068 | 1.0 | 22.1 | 0.45 | 2.53 |
| Low income | 0.012 | 0.068 | 1.5 | 33.2 | 0.94 | 5.33 |

Notes: Column (4) is calibrated so that expected global loss is 30% of GDP when the GDP's of different regions are weighted by their relative vulnerabilities in column (3). The willingnesses to pay for each region in columns (5) and (6) assumes a rate of relative risk aversion of 4, the probabilities in columns (1) and (2), and the expected losses in column (4).

**Table 4.10**
Summary of impacts in different sectors: impact of 2.5 degree warming (positive numbers are damages; negative numbers are benefits; impacts measured as percent of market GDPs)

| | Total [2.5 degree] | Agriculture | Other vulnerable market | Coastal | Health | Nonmarket time use | Settlements | Catastrophic impact [2.5 degree] | Catastrophic impact [6 degree] |
|---|---|---|---|---|---|---|---|---|---|
| United States | 0.45 | 0.06 | 0.00 | 0.11 | 0.02 | -0.28 | 0.10 | 0.44 | 2.97 |
| China | 0.22 | -0.37 | 0.13 | 0.07 | 0.09 | -0.26 | 0.05 | 0.52 | 3.51 |
| Japan | 0.50 | -0.46 | 0.00 | 0.56 | 0.02 | -0.31 | 0.25 | 0.45 | 3.04 |
| OECD Europe | 2.83 | 0.49 | 0.00 | 0.60 | 0.02 | -0.43 | 0.25 | 1.91 | 13.00 |
| Russia | -0.65 | -0.69 | -0.37 | 0.09 | 0.02 | -0.75 | 0.05 | 0.99 | 6.74 |
| India | 4.93 | 1.08 | 0.40 | 0.09 | 0.69 | 0.30 | 0.10 | 2.27 | 15.41 |
| Other high income | -0.39 | -0.95 | -0.31 | 0.16 | 0.02 | -0.35 | 0.10 | 0.94 | 6.39 |
| High-income OPEC | 1.95 | 0.00 | 0.91 | 0.06 | 0.23 | 0.24 | 0.05 | 0.46 | 3.14 |
| Eastern Europe | 0.71 | 0.46 | 0.00 | 0.01 | 0.02 | -0.36 | 0.10 | 0.47 | 3.23 |
| Middle income | 2.44 | 1.13 | 0.41 | 0.04 | 0.32 | -0.04 | 0.10 | 0.47 | 3.21 |
| Lower middle income | 1.81 | 0.04 | 0.29 | 0.09 | 0.32 | -0.04 | 0.10 | 1.01 | 6.86 |
| Africa | 3.91 | 0.05 | 0.09 | 0.02 | 3.00 | 0.25 | 0.10 | 0.39 | 2.68 |
| Low income | 2.64 | 0.04 | 0.46 | 0.09 | 0.66 | 0.20 | 0.10 | 1.09 | 7.44 |
| Global (a) | | | | | | | | | |
| Output-weighted | 1.50 | 0.13 | 0.05 | 0.32 | 0.10 | -0.29 | 0.17 | 1.02 | 6.94 |
| Population-weighted | 1.88 | 0.17 | 0.23 | 0.12 | 0.56 | -0.03 | 0.10 | 1.05 | 7.12 |

Note: (a) Output-weighted global average is weighted by projected output in 2100 from RICE base case. Population-weighted global average is weighted by population in 1995.

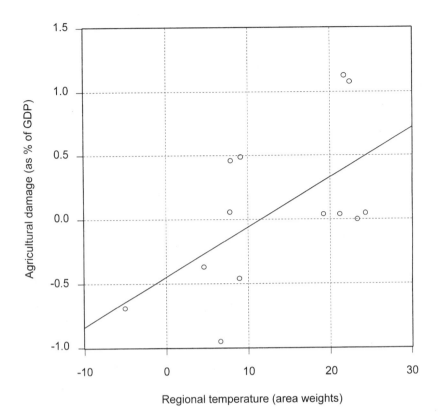

**Figure 4.1**
Agricultural damage function
Circles are estimated damages from table 4.4.

and temperature, which implies that countries with an initial climate cooler than 11.5°C will benefit while those that are warmer will be disadvantaged. For RICE-99, this relationship is used to project the relationship between agricultural damage and climate. These assumptions then yield an agricultural damage function for subregion $j$ of:

$$Q_{ag,j}(T) = \left\{ \alpha_{ag}^{\ 0} + \alpha_{ag}^{\ 1}(T + T^o_{\ j}) + \alpha_{ag}^{\ 2}(T + T^o_{\ j})^2 \right\}$$
$$- \left\{ \alpha_{ag}^{\ 0} + \alpha_{ag}^{\ 1}T^o_{\ j} + \alpha_{ag}^{\ 2}T^o_{\ j}^{\ 2} \right\} + \varepsilon_{ag,j} \tag{4.2}$$

where $\alpha_{ag}^{\ i}$ are the coefficients of the relationship, $T^o_{\ j}$ is subregional mean temperature in the absence of climate change, and $\varepsilon_{ag,j}$ is the error in the estimate for subregion $j$, chosen so that the impact for $T = 2.5$ equals the number in table 4.4. Because several subregions have very little solid information on subregional damages, we have assumed that

$\in_{ag,j} = 0$ for the following subregions: Eastern Europe, middle income, lower middle income, Africa, and low income.

**Other vulnerable market.**  Other market damages as estimated here show a strong relationship to subregional temperature, with the damage being zero at an initial subregional temperature of 12.3 degrees °C. The same methodology is followed as for agriculture for projecting the relationship for each subregion.

**Coastal.**  Coastal vulnerability is not related to subregional climate but to global processes. Following the work of Yohe, we assume a power relationship as follows:

$$Q_{coastal,j}(T) = \alpha_{coastal,j} \times [(T/2.5]^{1.5}. \tag{4.3}$$

The coefficient in this relationship is simply the estimated damage coefficient shown in table 4.10.

**Health.**  The estimates of the climatic impact on health are highly temperature dependent. To a first approximation, estimated health impacts are nil until subregional mean temperature reaches around 15°C, after which they rise sharply. To approximate this relationship, we estimate a semilogarithmic relationship between health damage and regional temperature. This relationship takes the form:

$$Q_{health,j}(T_j) = 0.002721 (T_j)^{0.2243}. \tag{4.4}$$

Figure 4.2 shows the estimates of the damage from a 2.5°C warming derived in the earlier section as triangles; the circles show the estimated damages for 2.5°C warming from the continuous damage function.

**Nonmarket amenity (time use).**  The estimates of impacts upon time use are based on a quadratic relationship between damage and subregional temperature. The methodology for calibration is the same as that for agriculture.

**Settlements.**  The estimates of damages for settlements are based upon the global change in temperature. The methodology for calibration is the same as that for coastal impacts.

**Catastrophic.**  The estimates of catastrophic damages in the last section provide estimates for both 2.5°C and 6°C changes. The damage

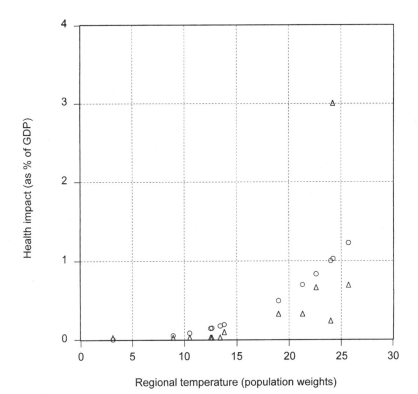

**Figure 4.2**
Health damages from model and Murray-Lopez study
Triangles are derived as described in text. Circles are estimated from nonlinear least
squares as discussed in text.

function is assumed to be linear for temperature changes up to 3°C and
then follow a power function for temperatures above 3°C.

## Calibration of the RICE-99 Damage Function

To go from the impact indices discussed above to the damage functions
$D_j(t)$ in RICE-99, these steps were followed:

1. Calculate the regional impacts for 2.5°C and 6°C, these being $\theta_{ij}[2.5,$
$y_j(2100)]$ and $\theta_{ij}[6, y_j(2100)]$. $y_j(2100)$ is taken from a run of RICE-98.

2. Sum these across categories to create overall impact indices for each
subregion to get the subregional aggregates, $\theta_j[2.5, y_j(2100)]$ and $\theta_j[6,$
$y_j(2100)]$.

3. Solve a system of two quadratic equations for each region to obtain the damage coefficients for the quadratic damage function for each RICE-99 region. The equations go through the points for temperature change of 0°C, 2.5°C, and 6°C. They thus represent the damage estimates in the first part of this chapter exactly and interpolate with a quadratic equation between 0°C and 6°C. Because none of these calculations exceed the 6°C increase, there is no need to extrapolate outside the range of estimates provided here. Note as well that for those RICE-99 regions that contain more than one of the thirteen subregions, we weight the subregions by GDPs.

## Major Results and Conclusions

The global damage function in RICE-99 is shown in figure 4.3. The two curves show the weighted sum of the regional damage functions at each temperature, where the weights are 1995 population and projected 2100 regional outputs. The results for individual subregions are shown in table 4.10. The subregional damage functions are shown in figure 4.4.

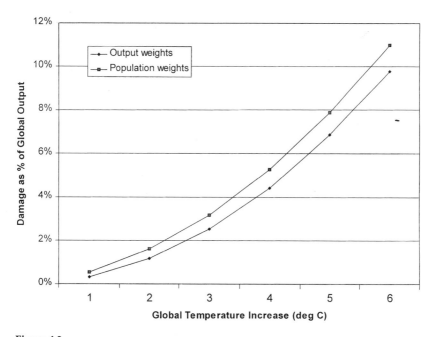

**Figure 4.3**
Global damage function

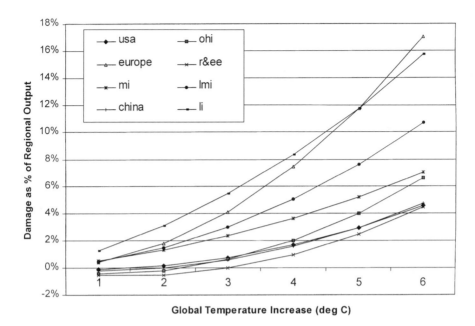

**Figure 4.4**
Regional damage functions

The results differ markedly by region. The impacts of a 2.5°C global warming range from a net benefit of 0.7 percent of output for Russia to a net damage of almost 5 percent of output for India. The global average impact of a 2.5°C global warming is estimated to be 1.5 percent of output using projected output weights and 1.9 percent of output using 1995 regional population weights.

One of the major surprises is that the impacts from global warming are likely to be quite modest for the next century. Current projections of RICE-99 in chapter 7 indicate that total warming in an uncontrolled environment will be slightly below 2.5°C around 2100. Our estimate is that damages at that time are likely to be around 1.9 percent of global income using 2100 output weights. The damages for the United States, Japan, Russia, and China are essentially zero over this time frame, assuming that catastrophic scenarios do not materialize. Europe, India, and many low-income regions, by contrast, appear vulnerable to significant damages over the next century.

The United States appears to be less vulnerable to climate change than many countries. This is the result of its relatively temperate climate, small dependence of its economy on climate, the positive

**Table 4.11**
Comparison of recent impact studies, United States: 2.5° C (billions of 1990 dollars; benefits are negative while damages are positive.)

|  | Fankhauser 1995 | Tol 1995 | Mendelsohn and Neumann 1999 | This study |
|---|---|---|---|---|
| Sector |  |  |  |  |
| *Market impacts* |  |  |  |  |
| Agriculture | 8 | 10 | −11 | 4 |
| Energy | 8 | na | 3 | 0 |
| Sea level | 9 | 9 | 0 | 6 |
| Timber | 1 | na | −3 | 0 |
| Water | 16 | na | 4 | 0 |
| Total market | 42 | 19 | −8 | 11 |
| *Nonmarket impacts* |  |  |  |  |
| Health, water quality, and human life | 19 | 37 | 6 | 1 |
| Migration | 1 | 1 | na | na |
| Human amenity, recreation and nonmarket time | na | 12 | −4 | −17 |
| Species loss | 8 | 5 | na | na |
| Human settlements | na | na | na | 6 |
| Extreme and catastrophic events | 0 | 0 | na | 25 |
| Total nonmarket | 28 | 56 | 2 | 17 |
| *Total (market and nonmarket sectors)* |  |  |  |  |
| Billions of 1990 $ | 70 | 74 | −7 | 28 |
| % of 1990 GDP | 1.3 | 1.5 | −0.1 | 0.5 |

Source: Table 4.1; Mendelsohn and Newmann 1999, pp. 319, 320; and the present study.
Note: na = not available or not estimated.

amenity value of a warmer climate in many parts of the United States, its advanced health system, and low vulnerability to catastrophic climate change. Table 4.11 compares the results of the present study with other recent studies for the United States.[16] The two most recent studies on the United States are largely in agreement that the economic impact of gradual climate change (that is, omitting catastrophic outcomes) is close to zero for a moderate (2.5° C) global warming.

16. The study by Mendelsohn and Neumann 1999 was published after the present study was completed and should be viewed as an independent evaluation.

Outside the United States, it is estimated that Russia and other high-income countries such as Canada will benefit slightly from a 2.5°C benchmark warming; the benefits to these regions come because of significant improvements in the agricultural sector as well as gains from nonmarket time use. At the other extreme, low-income regions—particularly India and Africa—and Europe appear to be quite vulnerable to climate change. The impact on India comes from its extreme vulnerability to climatic shifts because of the importance of monsoons on agriculture, the disamenity of increasing temperatures on nonmarket time use, and the potential for adverse health impacts. For Africa, much of the vulnerability comes from potential health impacts of global warming. Europe appears to be the most vulnerable of high-income regions because of the potential of catastrophic climate change due to shifts in ocean currents as well as significant coastal and agricultural impacts.

Estimates here indicate that for most countries the market impacts are likely to be relatively small; the major concerns are the potentially catastrophic impacts. As table 4.10 shows, the catastrophic costs are estimated to be twice as large as all other impacts combined for a 2.5°C warming. Similarly, catastrophic damages are estimated to dominate impacts for higher temperature increases. Because the estimated catastrophic impacts are so uncertain, this implies great uncertainty about the overall impacts.

A word of caution is necessary before closing. It must be emphasized that attempts to estimate the impacts of climate change continue to be highly speculative. Outside of agriculture and sea-level rise for a small number of countries, the number of scholarly studies of the economic impacts of climate change remains small. Estimates of the regional climatic impacts of global warming are still inconsistent across different climate models, and economic studies have made little progress in estimating impacts, particularly in low-income countries. Much more work is needed to improve understanding of the impacts of climate change.

# 5                  The DICE-99 Model

Earlier chapters discussed the development of RICE-99. For many purposes, particularly when the regional details are not essential, it is convenient to have a simplified version of the model. With this goal in mind, a globally aggregated model, which is called DICE-99, has been developed.

While losing the regional detail of RICE-99, DICE-99 has several advantages. It is more useful for understanding the basic structure of economic policy issues posed by greenhouse warming because it is small enough for researchers to understand the individual linkages in an intuitive way. It is more easily modified because the number of parameters is far smaller. It is much faster, so that alternative experiments can be tested more easily. And it can be run much further into the future so that the implications of alternative time horizons, discounting assumptions, and carbon or climate models can be more easily traced out. Researchers or policymakers who are interested in having an intuitive understanding of the economics of global warming are well-advised to begin with DICE before tackling more opaque and computationally demanding models such as RICE or other large-scale models.

## Model Structure

The basic structure of DICE-99 parallels RICE-99 in most sectors. The equations of DICE-99 are provided in appendix B, while the computer code for DICE-99 is provided in appendix E. The major difference between the two models lies in the production sector, where the reduced-form approach of the original DICE-94 model has been retained. More specifically, the major elements of DICE-99 include the following:

1. The geophysical sectors in DICE-99 are identical to RICE-99. The carbon cycle, radiative forcing, and climate equations are globally aggregated, so there is no reason to differentiate between the RICE and DICE models in these segments.

2. The treatment of the pure rate of time preference is identical for the RICE and DICE models.

3. The modeling structure for population, economy-wide technological change, labor inputs, investment, and the capital stock are identical. The only difference is that DICE represents the globally aggregated magnitudes, while RICE considers each of these variables separately for each region.

4. The damage equation takes the same form in the RICE and DICE models.

5. The major difference between the two models lies in the treatment of production and energy. As described above, RICE-99 introduces a more complete model of the energy sector, with carbon-energy entering as an intermediate input in the production function; DICE-99 uses a simplified reduced-form treatment of production. The DICE model has a Cobb-Douglas production function in labor, capital, and exogenous technological change. Base industrial carbon emissions are given by the product of a carbon-intensity factor times output, $\sigma(t)$, which is the ratio of uncontrolled industrial $CO_2$ emissions to global output. In runs where $CO_2$ emissions are controlled, emissions are reduced by the control rate, $\mu(t)$; controlled emissions are equal to base emissions times one minus the control rate. Net output is then gross output times a factor that is a function of the emissions control rate and the damages from climate change.

6. The final difference between the two models is the energy supply sector. In RICE-99, we explicitly model the exhaustion of carbon-energy. In DICE-99, all constraints on total use of carbon fuels are removed. In other words, there are no scarcity constraints on cumulative carbon-energy use in DICE-99.[1]

---

1. Carbon scarcity cannot be easily introduced in the DICE framework because of the reduced-form treatment of emissions reductions. Substitution away from carbon fuels occurs only when the emissions-reduction variable ($\mu$) is allowed to take nonzero values. The base case constrains $\mu$ to be zero. Scarcity-induced (as opposed to climate-policy-induced) substitution away from carbon fuels cannot be incorporated easily in this framework. Test runs using the standard version of DICE-99 indicate that there is no substantial impact of scarcity of carbon fuels for over 100 years. More precisely, if carbon scarcity similar to that in RICE-99 is introduced in DICE, a small Hotelling rent on carbon

In summary, DICE-99 is very similar in structure to the original DICE model. Those who are familiar with the earlier model will find that the new DICE version requires little learning time and is easy to use and manipulate. The new model is available in both a GAMS version and an EXCEL spreadsheet version, which means that the model can be used with inexpensive and widely available software. The major change from the previous version is recalibration to fit the new findings of the larger and more accurate economic structure of RICE-99. In addition, minor changes are made in the specification of certain parts of the model.

**Calibration**

DICE-99 was calibrated so that the output from its base run and its optimal run would fit the corresponding runs of RICE-99. The following explains the approach to calibration. Because of the highly divergent patterns of regional development, the fit between disaggregated RICE and aggregated DICE was imperfect, so the RICE and DICE models provide different projections for some variables.

The base case is described in chapter 2, the fifth section. The optimal run in DICE maximizes global utility subject to the major economic and physical constraints (a full listing is provided in appendix B). The optimal run in RICE-99 found a time path of carbon emissions that is Pareto optimal. Further description of the base and optimal runs can be found in chapter 6 and chapter 7, the second section.

Population, carbon intensity, the initial capital-output ratio, and economy-wide technological change are exogenous variables in DICE. These were set so that the paths of global population, global output, global emissions, $CO_2$ concentrations, and global temperature for the base run of DICE-99 matched those for the base run of RICE-99 over the first thirteen periods (130 years).

A more detailed discussion of the calibration procedure is now given. Population was calibrated so that it closely matched the path of aggregate population in RICE-99. Next the initial capital stock was calibrated so that the initial DICE-99 real return on capital was equal to the

---

fuels will come into play. The calculated Hotelling rent on carbon fuels is about $0.50 per ton carbon in 2000 and around $26 per ton carbon in 2100. This is suppressed in DICE, leading to slightly higher emissions and climate change. The difference in global mean temperature between the carbon-scarce and carbon-superabundant runs, however, is extremely small for two centuries—0.018°C in 2100 and 0.13°C in 2200.

output-weighted average real rate of return across regions in RICE-99.
Next, the initial level, initial growth rate, and decline in the growth rate
of total factor productivity in DICE-99 were set to match the initial level
of output, the average output level in the first four periods, and the
average output level in the first eleven periods in RICE-99.

Next, the level of the initial carbon-output ratio, $\sigma(0)$, was set so that
emissions in the first period matched actual emissions. Then, the
decline in $\sigma(t)$ was set so that the path of global temperature in DICE-
99 tracked RICE-99.

Table 5.1 and figure 5.1 show the percentage error of DICE-99 relative to RICE-99 for the important variables in the base run. As can be
seen, the average error for the important climatic variables is less than
2 percent over the next century.

The next step was to have the optimal run of DICE-99 match the
optimal run of RICE-99. For this step, the parameters of the damage
and emissions-control cost functions of DICE were adjusted. More
precisely, note that the cost of abatement function in DICE takes the
form $Cost(t)/Y(t) = [1 - b_1(t)\mu(t)^{b_2}]$, where $\mu(t)$ is the emissions-control

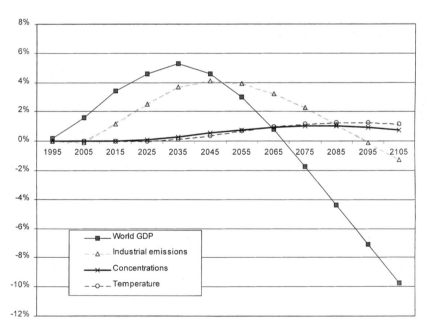

**Figure 5.1**
Calibration error in DICE reference case

**Table 5.1**
Comparison of RICE-99 and DICE-99 results, reference case (ratio of calculation for DICE-99 to RICE-99)

| | 1995 | 2005 | 2015 | 2025 | 2035 | 2045 | 2055 | 2065 | 2075 | 2085 | 2095 | 2105 |
|---|---|---|---|---|---|---|---|---|---|---|---|---|
| World GDP | 1.002 | 1.016 | 1.034 | 1.046 | 1.053 | 1.046 | 1.030 | 1.008 | 0.983 | 0.956 | 0.929 | 0.902 |
| World GDP/capita | 1.002 | 1.017 | 1.035 | 1.047 | 1.054 | 1.047 | 1.031 | 1.009 | 0.983 | 0.956 | 0.928 | 0.901 |
| World population | 1.000 | 0.999 | 0.999 | 0.998 | 0.999 | 0.999 | 0.999 | 0.999 | 0.999 | 1.000 | 1.001 | 1.001 |
| Industrial emissions | 1.000 | 0.999 | 1.012 | 1.026 | 1.037 | 1.041 | 1.039 | 1.033 | 1.023 | 1.011 | 0.999 | 0.987 |
| Total emissions | 1.000 | 0.999 | 1.011 | 1.023 | 1.034 | 1.038 | 1.037 | 1.031 | 1.022 | 1.011 | 0.999 | 0.988 |
| Industrial $CO_2$/output ratio | 0.999 | 0.985 | 0.980 | 0.983 | 0.988 | 0.999 | 1.014 | 1.030 | 1.047 | 1.064 | 1.082 | 1.102 |
| Concentrations | 1.000 | 1.000 | 1.000 | 1.001 | 1.003 | 1.006 | 1.008 | 1.009 | 1.010 | 1.010 | 1.009 | 1.008 |
| Global temperature | 1.000 | 1.000 | 1.000 | 1.000 | 1.001 | 1.004 | 1.007 | 1.010 | 1.012 | 1.013 | 1.013 | 1.012 |
| Cumulative total emissions | 1.000 | 1.000 | 1.004 | 1.009 | 1.015 | 1.019 | 1.022 | 1.023 | 1.023 | 1.022 | 1.019 | 1.016 |
| Average concentration | 1.000 | 1.000 | 1.000 | 1.000 | 1.001 | 1.002 | 1.003 | 1.004 | 1.005 | 1.005 | 1.006 | 1.006 |
| Average temperature | 1.000 | 1.000 | 1.000 | 1.000 | 1.000 | 1.001 | 1.003 | 1.004 | 1.006 | 1.007 | 1.008 | 1.008 |

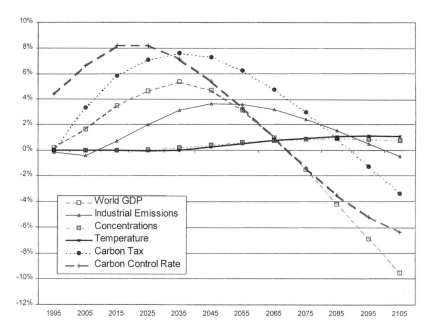

**Figure 5.2**
Calibration error in DICE optimal case

rate.[2] The coefficients $b_1(t)$ and $b_2$ and those of the quadratic damage
function were set so that the optimal carbon tax and emissions control
rates in DICE-99 matched the projections of these variables in the
optimal run of RICE-99.

Table 5.2 and figure 5.2 show the calibration errors for the optimal
run. The average errors for the first twelve periods (DICE-99 relative
to RICE-99) are 1.6 percent for industrial emissions, 0.4 percent for concentrations, 0.5 percent for temperature increase, 3.5 percent for the
carbon tax, and 2.3 percent for the emissions-control rate. In short, the
calibrated DICE-99 model is a faithful reflection of RICE-99.

A final word will be helpful for those contemplating whether to use
the RICE or DICE model as research tools. DICE is much easier to use
and runs much more quickly. The two models track closely for the first
150 years, after which numerical approximations and the shorter time
horizon become a problem in RICE. For looking at longer-run trade-
offs, particularly those that do not involve regional analyses, DICE is

2. If $E_b(t)$ is industrial emissions in the baseline, the emissions control rate for period $t$,
$\mu(t)$, is $[E_b(t)-E(t)]/E_b(t)$.

**Table 5.2**
Comparison of RICE-99 and DICE-99 results, optimal case (ratio of calculation for DICE-99 to RICE-99)

| | 1995 | 2005 | 2015 | 2025 | 2035 | 2045 | 2055 | 2065 | 2075 | 2085 | 2095 | 2105 |
|---|---|---|---|---|---|---|---|---|---|---|---|---|
| Total emissions | 0.999 | 0.996 | 1.006 | 1.018 | 1.029 | 1.034 | 1.034 | 1.030 | 1.023 | 1.015 | 1.005 | 0.995 |
| $CO_2$/output ratio | 0.997 | 0.981 | 0.975 | 0.977 | 0.982 | 0.993 | 1.009 | 1.027 | 1.046 | 1.066 | 1.086 | 1.107 |
| Industrial emissions | 0.998 | 0.996 | 1.007 | 1.020 | 1.032 | 1.036 | 1.036 | 1.032 | 1.024 | 1.015 | 1.005 | 0.995 |
| Cumulative total emissions | 0.999 | 0.997 | 1.001 | 1.005 | 1.011 | 1.015 | 1.018 | 1.020 | 1.020 | 1.020 | 1.018 | 1.016 |
| Carbon control rate | 1.044 | 1.066 | 1.082 | 1.082 | 1.072 | 1.055 | 1.033 | 1.010 | 0.986 | 0.965 | 0.948 | 0.936 |
| Carbon tax | 1.000 | 1.033 | 1.059 | 1.071 | 1.076 | 1.073 | 1.063 | 1.048 | 1.030 | 1.009 | 0.987 | 0.966 |
| Concentrations | 1.000 | 1.000 | 1.000 | 1.000 | 1.002 | 1.004 | 1.006 | 1.008 | 1.009 | 1.009 | 1.009 | 1.008 |
| Global temperature | 1.000 | 1.000 | 1.000 | 0.999 | 1.000 | 1.002 | 1.005 | 1.007 | 1.009 | 1.011 | 1.011 | 1.011 |
| Average concentration | 1.000 | 1.000 | 1.000 | 1.000 | 1.000 | 1.001 | 1.002 | 1.003 | 1.003 | 1.004 | 1.005 | 1.005 |
| Average temperature | 1.000 | 1.000 | 1.000 | 1.000 | 1.000 | 1.000 | 1.001 | 1.003 | 1.004 | 1.005 | 1.006 | 1.007 |

a more accurate instrument and much easier to use. Problems can arise in either model when it is run too far outside the area for which it was designed and calibrated. For example, in DICE, increasing economic growth rates, population, or carbon intensities may increase total use of carbon fuels well beyond current estimates of availability. By contrast, RICE-99 contains an upward-sloping supply curve for fossil fuels, and this constraint prevents excessive cumulative emissions. Caution should be taken to ensure that analyses using the models do not violate implicit assumptions used to simplify the model.

# 6      Computational Procedures

## Computer Programs for RICE and DICE

The RICE-99 and DICE-99 models have been programmed in two different versions. The most accessible version of the models, available on the Internet, are programmed using Microsoft EXCEL (97 or later). In addition, the models have been programmed with a widely used mathematical programming language, GAMS (version 2.50 or later). The current version of the GAMS program has been tested only on the solver MINOS5. The GAMS programs are provided in appendixes D and E of this book. The EXCEL spreadsheet and GAMS programs, along with documentation, are available on the internet at *http://www.econ.yale.edu/~nordhaus/homepage/homepage.htm.*

The EXCEL and GAMS versions are identical for the DICE-99 model. The implementations of the equilibrium, however, are slightly different in the two versions for the RICE-99 model. This chapter describes the solution procedures for the models.

## Solution Approach in EXCEL—RICE-99

The discussion begins by describing the solution concept for RICE-99 in the EXCEL approach. In this version, different regions are investing and consuming to optimize their regional social welfare functions without the possibilities of international trade. This equilibrium produces different discount rates on goods (real interest rates) across regions. The externality from global warming is then incorporated by estimating the world's willingness to pay as the sum of the damages across different regions valued at the different regions' discount rates. In those cases where emissions trading occurs, each region will optimize its carbon emissions at the world-market carbon tax. Each region's net permit revenue is the carbon tax times net permit sale (i.e., permit allocation minus emissions). GDP then includes net permit revenues. The calculation of the global carbon tax, permit revenues, outputs, and savings rates are all done simultaneously and consistently. For revenue-neutral permit allocations, net permit revenues are set equal to zero. This has the same effect as if we allocated each region's permits to be equal to its equilibrium emissions, but simplifies the calculation by allowing us to avoid the iterations necessary to find the revenue-neutral permit allocations that are consistent with the time path of emissions.

The rows in the EXCEL spreadsheet can be classified as exogenous or endogenous. The exogenous rows consist of numbers that have been entered by the user, and the endogenous rows are formulas. The exogenous rows can be further divided into parameters and policy variables. The policy variables are those that can be determined through the choice of a policymaker rather than by nature: regional savings rates and carbon taxes. The Hotelling rent is in this category although it is a market price. Industrial carbon emissions and carbon-energy are endogenous variables in the spreadsheet. Each region's carbon-energy and industrial emissions satisfy a first-order condition for output maximization (equation 3.3b); the price of carbon-energy, which includes the carbon tax, enters into the formula for emissions. The parameter rows, which contain values for population, total factor productivity, the rate of decarbonization, nonindustrial carbon emissions, and production function parameters, would generally be viewed as outside the control of a policymaker.

The model is solved in EXCEL by finding the values of the policy variables that meet the user's desired conditions. Table 6.1 displays the conditions for solution of four policies considered in chapter 7 (policies 1, 2, 5, and 6 in table 7.1). Column (3) displays the mathematical condition in EXCEL that corresponds to the conceptual condition in column (2). The solutions in chapter 7 are found by searching for the values of the policy variable in column (4) that satisfy the condition in (3). The searching is usually done via the "goal seek" command. To ensure that all conditions for a solution are met in all periods, iteration is employed in the solution algorithm. Macros have been programmed into the EXCEL spreadsheet so that the cases considered in this book can be solved with minimal effort.

All the cases considered in chapters 7 and 8 are minor variants of the four policies found in table 6.1, and extension of the solution method in table 6.1 is straightforward. The user always has the option of explicitly specifying the policy variables in the EXCEL spreadsheet.

The solution approach followed in the EXCEL spreadsheet is made possible because the RICE-99 model structure is so simple that the mathematical conditions for a solution can be derived analytically. It is possible to imagine cases where this cannot be done. One such case is the concentrations target case (number 4, table 6.1). Here, however, we can modify the problem by making damages [$D_j(t)$ in equation (2.16)] a function of concentrations that rises sharply near the concentrations target; solving as in the Pareto optimal case (number 2, table 6.1) will then provide a reasonably accurate solution.

### Solution Approach in GAMS—RICE-99

An alternative approach to solving the model is to maximize a social welfare function subject to the equations of the model using an optimization program such as GAMS. Our approach is a variant of the Negishi method of solving for a general equilibrium. The original approach comes from T. Negishi 1960 and was discussed in Nordhaus and Yang 1996.

Negishi's idea is that a competitive equilibrium can be found by maximizing a social welfare function—the weighted sum of the utility functions of each of the economy's agents—which has the appropriate set of welfare weights. Here, the agents are regions. The weights are ones that induce zero transfers among regions in the solution path, or that equalize the shadow prices on the individual agent's budget constraints.

In the pure Negishi approach, the social welfare function is:

$$SWF_{PURE} = \sum_J \psi_J W_J, \tag{6.1}$$

where $SWF_{PURE}$ is the social welfare function in the pure Negishi method, the $\psi_J$ are welfare weights, and $W_J$ is from equation (2.1).

The original or pure approach cannot be applied to our problem. Since we do not allow intertemporal trade, each region is forced to consume and invest solely out of its own output, and it therefore faces multiple constraints on its expenditures—one for each period—rather than a single constraint on the present value of its expenditures. A set of time-invariant social welfare weights does not exist that equates the social shadow prices of the budget constraints in each period. In other

**Table 6.1**
The basic policies of the RICE model

| (1) Case | (2) Conditions for solution | (3) Mathematical condition | (4) Control variable |
|---|---|---|---|
| 1. Base | 1. Savings rates optimized in each region. | Rate of return on capital determined by regional utility discount rate | Regional savings rate |
| | 2. Hotelling rent equals scarcity rent. | Hotelling rent equals discounted value of effect of marginal extraction on price | Hotelling rent |
| | 3. Externality ignored in choice of emissions | Permit price (carbon tax) equals zero | World permit price or carbon tax |
| | 4. Industrial emissions satisfy market equilibrium with carbon tax (permit price) of zero. | First-order condition for emissions (equation 3.3a) | Formula for industrial emissions ensures that first-order conditions are met. |
| 2. Pareto optimum | 1 and 2 from base case | | |
| | 3. Samuelson condition for public goods (marginal cost equals sum of regional marginal benefits). | Uniform carbon tax (permit price) equals world marginal willingness to pay for carbon abatement. | World permit price or carbon tax |
| | 4. Industrial emissions satisfy market equilibrium with optimal carbon tax. | First-order condition for emissions (equation 3.3a) | Formula for industrial emissions ensures that first-order condition is met. |
| | 5. Marginal cost of carbon abatement is the same across countries | Derivative of GDP with respect to industrial emissions is the same in each region. | Ensured by fact that each region faces same carbon tax or permit price. |

| | | |
|---|---|---|
| 3. Global emissions limit | 1 and 2 from base case | |
| | 3. Global emissions limited by policy constraint | Demand for emissions (according to formula) equals world emissions limit | World permit price or carbon tax |
| | 4. Industrial emissions satisfy market equilibrium with nonzero carbon tax. | First-order condition for emissions (equation 3.3a) | Formula for industrial emissions ensures that this is met. |
| | 5. Marginal cost of abatement is the same across countries. | Derivative of GDP with respect to industrial emissions is the same in each region. | Ensured by fact that each region faces same carbon tax or permit price. |
| 4. Concentrations limit | 1 and 2 from base case | |
| | 3. Emissions are Pareto optimal subject to limit on $CO_2$ concentrations. | Same as #3 in Pareto optimum, with a highly nonlinear damage function near the concentration limit | World permit price or carbon tax |
| | 4. Industrial emissions satisfy market equilibrium with nonzero carbon tax (or permit price). | First-order condition for emissions (equation 3.3a) | Formula for industrial emissions ensures that this is met. |
| | 5. Marginal cost of abatement is the same across countries. | Derivative of GDP with respect to industrial emissions is the same in each region. | Ensured by fact that each region faces same carbon tax or permit price. |

words, without intertemporal trade, discount rates will differ across regions. Hence, if a set of time-invariant welfare weights equalizes the marginal social utility of consumption across regions in one period, the marginal social utilities will diverge in future periods.

The approach here uses time-varying weights. The social welfare function is:

$$SWF_{TVW} = \sum_t \sum_J \psi_J(t)U[c_J(t), L_J(t)], \tag{6.2}$$

where $SWF_{TVW}$ is the social welfare function in the Negishi method with time-varying weights $(TVW)$, $\psi_J(t)$ are time-varying welfare weights, and $U$ is from equation (2.1). The welfare weights were chosen so that the shadow prices on the period-specific budget constraints—the social marginal utilities of income—are the same across regions in each period at the social optimum.

With the welfare weights chosen in such a manner, one gets:

$$dSWF_{TVW}/dC_J(t) = WR(t) \tag{6.3}$$

at the solution. Equation (6.3) states that the social discount factor (or social discount rates) is the same for all regions.

Note that there is one degree of freedom in choosing the world discount rate, $WR(t)$; equating social marginal utilities in each period requires that the discount factor be equal across regions but does not pin the discount factor down. We choose $WR(t)$ so that it equals the output-weighted average of the individual-region discount factors. (This is close to setting it equal to the capital-weighted average of discount factors.) To do this, we employ an iterative search algorithm in which equation (6.4) is first maximized with an arbitrary $WR(t)$; we then adjust $WR(t)$ and maximize (6.4) again; and follow this procedure until the algorithm has converged. Since the individual-region discount factors are endogenous, this way of choosing $WR(t)$ has the effect of making the social discount factor in (6.4) endogenous. As society's future consumption changes relative to present consumption, changes in the relative valuation of future to present consumption will affect the social planner's decisions.

To vary policy in GAMS, we simply vary the constraints. Unlike the spreadsheet approach, emissions are treated as a control variable in GAMS, while the carbon tax and the Hotelling rent are calculated as output of the program.

To solve the base case in GAMS, we optimize under the assumption that the damage coefficients are exogenous; they are set to a level that is consistent with the market level of industrial emissions. To solve the optimal case, the constraints are all the equations in chapter 2. To solve case 3 in table 6.1, we add the constraint that global industrial emissions not exceed a certain level. In the concentrations target case, we instead constrain concentrations of carbon in the atmosphere to remain below a certain level. The Negishi method can be applied easily to other cases by adding to or modifying the GAMS program.

Using the TVW variant of the Negishi method, some of the conditions for a solution in table 6.1 will not be met exactly. Because the social planner applies a weighted-average discount rate to all regions, savings rates will not be optimized. Hence, in calculating the scarcity and environmental values of carbon abatement, future benefits will be discounted at too high a rate for some regions and too low a rate for others.

Experience indicates, however, that for the most part, the Negishi-TVW solution is acceptably close to one that meets the conditions in table 6.1. (See the comparisons in table 6.2.) In handling the trade-off between the present and far future, Negishi-TVW will be close to optimal because the calculation of future environmental damages requires summation across regions (so the underdiscounting cancels to

Table 6.2
Comparison between GAMS and EXCEL solutions

|  | GAMS output as % of EXCEL output | |
| --- | --- | --- |
|  | Base case | Pareto optimal case |
| Global mean temperature increase, 2105 | 101.6 | 101.1 |
| Global mean temperature increase, 2305 | 99.8 | 100.0 |
| Per capita income, ratio of 2105 to 1995, United States | 97.9 | 97.8 |
| Per capita income, ratio of 2105 to 1995, LI | 102.3 | 102.2 |
| World industrial carbon emissions, 2105 | 100.4 | 100.2 |
| Carbon tax, 2015 | NA | 124.0 |
| Carbon tax, 2105 | NA | 100.7 |
| Present value of consumption, world | 110.6 | 110.6 |
| Total economic impact of policy | NA | 118.3 |

some extent the overdiscounting) and because regional discount rates converge over time. As is discussed below, the Negishi-TVW has the advantage that no computational errors will creep in from emissions trading since the discount rates are equalized across regions.

In interpreting the EXCEL and GAMS approaches to solving RICE-99, it should be noted that there are differences in the global calculations of economic impacts because of differing treatment of trade and discounting. The EXCEL version uses regionally varying discount rates to discount the costs and benefits of policies to each region while the GAMS version uses a common discount rate for all regions. The differing discount rates will lead to problems in calculating the global impacts in the EXCEL version, particularly when trading occurs. The difficulty arises because the present values of unit of consumption in different regions are not equal. Hence, if a unit of output is transferred in a given future year from a high-income region like the United States (with a low discount rate) to a low-income region like China (with a high discount rate), the present value of world income will be lowered. The difficulties are most likely to cause problems in the cases that involve large flows in regimes with emissions trading. It should be noted that, because discount rates are equalized in the GAMS version, this problem does not arise there.

### DICE

Solving DICE is considerable easier because there is only one region— the world. To solve the model in GAMS, we simply maximize the welfare of this one region, varying the constraints to suit the problem. There is no question of which discount rate to use, so the problems that prevent finding an exact and consistent solution do not arise here. Solutions in the spreadsheet version are done in the same way as for RICE-99, but they take less time. With DICE, one also has the option of using the EXCEL solver to solve the problem the same way it is solved in GAMS, through welfare maximization, because a one-region model is often small enough for the solver to handle. Our experience with the solver is that it tends to converge short of the optimum.

### GAMS versus EXCEL

The alternative solution approaches offer different advantages and disadvantages.

## User Friendliness

The EXCEL spreadsheet is more user friendly. It lays out the full model structure before the modeler, making it very easy to understand what is going on and to see the consequences of changes in parameters or control variables. The user can simply change a parameter, tax rate, and so forth, and instantly see the results of the change. With GAMS, to see the consequences of a policy change, the user needs to run the program again, and the nature of the GAMS program is that it is easy to introduce unexpected bugs even with minor changes in the program. Although it is possible to put the output of the GAMS program into a form that lays out the model structure, this requires cutting and pasting each time the program is run.

## Model Output

The main EXCEL workbook itself makes a wide array of output available to the user, and output spreadsheets, linked to the main workbook, allow the user to quickly create a variety of tables, graphs, and additional variables from the model output. Additional variables that the modeler would like can be created on the spreadsheet with a few seconds' work. With the GAMS program, one can program it to produce the output one wants, but it requires some experience to learn what one needs; to see the results of changes, one must go back and run the program again. With some work, the user can automate the process by pasting GAMS output into a spreadsheet or reading it into a program, but this process is much more involved than working with the EXCEL spreadsheets.

## Accuracy of Solution

In the RICE-99 and DICE-99 models, for the experiments we have considered, both EXCEL and GAMS find exact solutions (within the tolerance limits of the programs) for the problems that are solved, although each has the shortcomings discussed above that are associated with the discount rates. The one case where the accuracy of GAMS is clearly superior to that of the EXCEL spreadsheet is the concentrations limit in DICE.

*Solution Time*

Significant differences in solution time between the two programs have not been found, assuming the user employs reasonable starting values in EXCEL. In our experience, solving the full RICE-99 model takes from 15 to 30 minutes on a 500-MHZ machine, depending on the case being solved.

If the user needs only a partial solution (for example, is satisfied to keep the savings rates constant but needs to find a Pareto optimal carbon tax and Hotelling rent given those rates), then the EXCEL spreadsheet can offer substantial advantages over the GAMS program because the user has the option of solving only the subproblem of interest. Finding the optimal savings rates takes by far the longest time of all of the subproblems in the EXCEL spreadsheet; yet, the savings rates for one case turn out to be nearly indistinguishable from the optimal rates in other cases. So, since the user can usually dispense with the most time-consuming part of the solution, the EXCEL spreadsheet in practice turns out to be the faster option. For cases 1 through 4 in table 6.1 when savings rates are held constant in the EXCEL worksheet, solution times range from 1 to 15 minutes.

If the user has an explicit schedule of savings rates and carbon taxes, the EXCEL spreadsheet offers near-instantaneous results once the control variables have been entered.

*Summary Advice for Users*

Users may want to change the RICE or DICE models to test alternative structures, assumptions, or policies. Changing the DICE model has proven relatively easy in either the EXCEL or GAMS versions. Adapting the RICE model generally is more difficult. Which version should be adapted depends upon the type of change and the programming background of the user. The following are guidelines for potential users:

1. Students and first-time modelers will find the EXCEL versions easier tools to use.

2. It is easy to test the results of different parameter values in EXCEL, whereas changing parameter values in GAMS requires reprogramming and making new runs.

3. Minor changes in the model structure are a close call. They involve adding to or changing the computer code in GAMS, while in EXCEL they involve entering formulas, changing formulas, adding or deleting spreadsheet rows, and changing the solution macros. Since there are eight regions in the model, each of whose structure may need to be changed, since the macros that solve the model will often need to be changed, especially if rows are added to or deleted from the spreadsheet, and since the code underlying the spreadsheet is less transparent, reprogramming the spreadsheet structure is likely to be more difficult than reprogramming the structure in the GAMS program.

4. Major changes in the model structure or examining alternative policies will change the equilibrium conditions for the model solution or their mathematical interpretation. In such cases, the GAMS program is generally the route to follow. Reprogramming the EXCEL spreadsheet to solve the model may be difficult or impossible, especially since it requires the user to be able to write down the conditions for a solution analytically.

# II Policy Applications of the RICE Model

# 7             Efficient Climate-Change Policies

The DICE-99 and RICE-99 models were laid out in the first part of this volume. The balance of the study applies these models to major issues of climate-change policy. The present chapter identifies a number of alternative approaches to climate change policy and investigates the relative efficiency of these alternatives. The next chapter then analyzes the current approach to climate-change policy—the Kyoto Protocol.

**Alternative Approaches to Climate-Change Policy**

This book uses integrated assessment (IA) modeling to assess the economic and environmental impacts of alternative approaches to climate-change policy. The advantage of using IA models is that the entire system can be analyzed simultaneously; that is, the impact of alternative policies on the environment and the economy can be analyzed as a package. This allows one to understand the tradeoffs involved in a more precise fashion.

Although there is a bewildering array of potential approaches to greenhouse warming, we have organized them into eight major polices shown in table 7.1. These can be grouped into four general categories: do nothing (policy 1); variants on an optimal policy (policies 2 and 3); arbitrary limitations on environmental variables (policies 4 through 7); and a major technological breakthrough (policy 8). This chapter discusses the relative advantages and disadvantages of these different approaches.

It is desirable to design policies that are economically efficient so that the environmental objectives can be attained in a least cost manner. There are four kinds of efficiency standards that can be examined: how-efficiency, where-efficiency, when-efficiency, and why-efficiency.

**Table 7.1**
Alternative policies analyzed in RICE-99 and DICE-99 models

1. *No controls (baseline).* No policies taken to slow greenhouse warming.

2. *Optimal policy.* Emissions and carbon prices set at Pareto optimal levels.

3. *Ten-year delay of optimal policy.* Delays optimal policy for ten years.

4. *Stabilize emissions of high-income regions (Kyoto Protocol).* Annex I regions reduce their emissions 5 percent below 1990 levels forever, with trading allowed among Annex I regions.

5. *Stabilizing global emissions.* Stabilizes global emissions at 1990 levels.

6. *Concentrations stabilization.* Stabilizes concentrations at two times preindustrial levels.

7. *Climate stabilization.* Sets policies to limit temperature rise to (a) 2.5°C or (b) 1.5°C.

8. *Geoengineering.* Implements a geoengineering option that offsets greenhouse warming at no cost.

· *How-efficiency* denotes the use of efficient ways of achieving emissions reductions in a given year and region. The current study assumes that individual regions attain how-efficiency by domestic auctioning of emissions permits (or equivalently via uniform carbon taxes).

· *Where-efficiency* denotes allocating emissions reductions across regions to minimize the costs of attaining the global emissions target for a given year. A policy where the only trading bloc is the entire world, which is true for all the policies in table 7.1 except number 4, will be where-efficient,[1] whereas a policy such as the Kyoto Protocol, in which there is more than one trading bloc with limited trading, will forfeit some of the gains from trade.

· *When-efficiency* refers to an efficient allocation of emissions over time. A when-efficient policy seeks an emissions path that minimizes the present value of the cost of emissions reductions, subject to the policy's environmental goal and the allocation of emissions reductions across regions. Policies 2, 6, and 7 seek efficient timing of emissions reductions. Policies 4 and 5 specify an arbitrary time path of global emissions; since they do not attempt to optimize on timing, they are not when-efficient.

· Finally, *why-efficiency* refers to attaining the ultimate objective of a program, which is here taken to be a set of policies that balances the costs of abatement and benefits of damage reduction. The optimal program in policy 2 satisfies why-efficiency and can therefore be used

---

1. If the entire world faces the same carbon tax, then the marginal cost of emissions reduction will be the same in each region. If there are two trading blocs, the marginal cost of emissions reduction will be high in one region and low in the other.

as a benchmark for why-efficiency comparisons with other proposals. The environmental goals of policies 4, 5, 6, and 7 are chosen arbitrarily, so these policies are not why-efficient.

## Detailed Description of Different Policies

### No Controls (Baseline)

The first run is one in which no policies are taken to slow or reverse greenhouse warming. Individuals and firms would adapt to the changing climate, but governments are assumed to take no steps to curb greenhouse-gas emissions or to internalize the greenhouse externality. This policy is one that has been followed for the most part by nations through 1999.

### Optimal Policy

The second case solves for an economically efficient or "optimal" policy to slow climate change. More precisely, this run finds a Pareto optimal trajectory for the world carbon tax (and thus for global industrial emissions), one that balances current abatement costs against future environmental benefits of carbon abatement. Permits are allocated in a revenue-neutral way across countries (Recall that a revenue-neutral permit allocation grants each region permits equal to its emissions at the equilibrium carbon tax.)

The optimal case is where-efficient, when-efficient, and why-efficient. Where-efficiency is guaranteed by the fact that the entire world is one trading bloc. The when- and why-efficient carbon tax is found by setting it equal to the environmental shadow price of carbon.

It will be useful to provide a word of interpretation of the optimal case. This is not presented in the belief that an environmental pope will suddenly appear to provide infallible canons of policy that will be scrupulously followed by all. Rather, the optimal policy is provided as a benchmark for policies to determine how efficient or inefficient alternative approaches may be.

### Ten-Year Delay of Optimal Policy

This case is one that delays implementing the optimal policy for ten years. The policy might be rationalized as one that allows nations to calculate the costs and benefits of delaying implementing policies until knowledge about global warming is more secure. In this scenario, we assume that sufficient information is in hand so that nations begin to act optimally starting in the period 2000–09.

## Kyoto Protocol

Many current policy proposals deal with intermediate objectives like stabilizing emissions. For example, the Kyoto Protocol of December 1997 is designed to limit the emissions of Annex I countries (essentially, OECD countries plus Eastern Europe and most of the former Soviet Union). The Protocol states: "The Parties included in Annex I shall, individually or jointly, ensure that their aggregate anthropogenic carbon dioxide equivalent emissions of the greenhouse gases . . . do not exceed their assigned amounts, . . . with a view to reducing their overall emissions of such gases by at least 5 percent below 1990 levels in the commitment period 2008 to 2012." In other words, Annex I countries during the period 2008–2012 will reduce their combined emissions of greenhouse gases on average by 5 percent relative to 1990 levels.

There are a number of ways of implementing the Kyoto Protocol. While the next chapter analyzes the Kyoto Protocol in detail, the present chapter looks only at the basic Kyoto Protocol design; the basic framework assumes that the Annex I emissions limit is constant indefinitely ("Kyoto forever") and allows trading of emissions rights among Annex I regions (Annex I trading). Annex I regions are allocated emissions permits as specified in the protocol. In RICE-99, Annex I is made up of USA, OHI, OECD Europe, and R&EE.[2]

Non-Annex I regions have unconstrained emissions in this case (non-Annex I carbon tax = 0).

## Stabilizing Global Emissions

The Kyoto Protocol targets the emissions only of high-income countries. A broadened Kyoto Protocol would include all countries. For this policy, we assume that global industrial emissions are limited to 1990 levels starting in 2005, and abatement is efficiently distributed around the world (i.e., the carbon tax is the same in all regions). As in policies

---

2. Annex I in RICE-99 does not correspond exactly with the actual Annex I, because OHI includes some countries that are not part of the actual Annex I. Emissions limits and permit allocations in case number 4 and in the cases in chapter 8 are scaled up appropriately.

2 and 3, permits are allocated so that net permit revenue is zero in all regions. In RICE-99, global reference industrial $CO_2$ emissions are estimated to be 6.19 GtC per year on average for 1990–99. We estimate that 1990 emissions were 0.916 times first-period emissions. Therefore, under this policy, global $CO_2$ industrial emissions are limited to 5.67 GtC per year. (These emissions exclude emissions from land-use changes, estimated to total 1.13 GtC per year in the 1990–1999 period.) Note that this policy is more stringent that the Kyoto Protocol, which limits emissions only of high-income countries but does not limit developing-country emissions.

### Concentrations Stabilization

One of the new approaches that has received considerable attention is to stabilize the concentrations of $CO_2$ in the atmosphere. This policy is motivated by two ideas. First, it is concentrations rather than emissions that will produce harmful and dangerous climate change. Second, $CO_2$ concentrations are closely related to $CO_2$ emissions, which are in principle under the control of policy. Concentrations were specifically identified under the U.N. Framework Convention on Climate Change, which states, "The ultimate objective of this Convention . . . is to achieve . . . stabilization of GHG concentrations in the atmosphere at a level that would prevent dangerous anthropogenic interference with the climate system."[3] Although no dangerous level has been established, some scientists believe that a prudent policy would be to limit $CO_2$ concentrations to two times their preindustrial levels. This policy is usually taken to be a threshold of 560 parts per million of $CO_2$, or about 1190 GtC carbon in the atmosphere.

In policy 6 we limit atmospheric concentrations to 1190 GtC or less and solve for a Pareto optimal carbon tax trajectory subject to this constraint. As in polices 2, 3, and 5, the entire world is one trading bloc with revenue-neutral allocation of permits.

### Climate Stabilization

A more ambitious approach involves taking steps to slow or stabilize the increase in global temperature so as to prevent major ecological impacts and other damage. This approach is particularly interesting because it focuses on an objective that is closer to the area of actual

---

3. Article 2.

concern—climate change—as opposed to most other policies, such as emissions or concentrations limits, which focus on intermediate variables of no intrinsic concern.

There have been a number of proposals for setting "tolerable windows" on climate change.[4] In policy 7a we limit the global mean temperature rise to 2.5°C. This is the IPCC's central estimate for the equilibrium temperature increase associated with a doubling of atmospheric carbon dioxide (IPCC 1996a). In the RICE-99 base case, this increase is reached in the first decade of the twenty-second century. In policy 7b we limit the temperature increase to 1.5°C. In both cases, we solve for cost-minimizing emissions trajectory subject to temperature remaining below the limit, and again we assume that the plan is implemented through harmonized carbon taxes or a revenue-neutral permit allocation. Additional proposals have been made to limit the rate of change of temperature. In practice, the proposed rate of change constraints do not bind for the two limits investigated here, so the rate of change constraints have been omitted.

Policies 2, 3, 5, 6, and 7 allocate emissions permits so that each region receives its own permit payments as revenues. As explained in chapter 2, the fifth section, these policies can be thought of as uniform and harmonized carbon taxation with lump-sum recycling of revenues within regions.

*Geoengineering*

A radical technological option would be geoengineering, which involves large-scale engineering to offset the warming effect of greenhouse gases. Such options include injecting particles into the atmosphere to increase the backscattering of sunlight and stimulating absorption of carbon in the oceans. The most careful survey of this approach by the 1992 report of the U.S. National Academy concluded, "Perhaps one of the surprises of this analysis is the relatively low costs at which some of the geoengineering options might be implemented."[5]

---

4. For a recent discussion, see Toth et al. 1998, which also calculates emissions trajectories that would keep climate safely beneath a temperature trajectory that might trigger changes in the thermohaline circulation. All runs of RICE-99 are well below the trigger trajectory.

5. National Academy of Sciences 1992, p. 460. The National Academy report describes a number of options that provide the theoretical capability of unlimited offsets to the radiative effects of GHGs at a cost of less than $1 per ton C (see National Academy of Sciences 1992, chapter 28).

Calculations here assume that geoengineering is costless. This is based on current findings which indicate that several geoengineering options are available that would cost less than $10 per ton carbon or have the globally averaged radiative effect of reducing emissions by one ton of carbon. It should be emphasized that many ecologists and environmentalists have grave reservations about the environmental impacts of the geoengineering options. Moreover, the climatic impacts have been insufficiently studied. Nonetheless, because of the high cost of other mitigation strategies, this scenario is useful as a benchmark to determine the overall economic impact of greenhouse warming and of policies to combat warming.

## Major Results

The results for the DICE-99 and RICE-99 models in this section have been obtained using the EXCEL versions.

### Overall Results

First, the overall results are summarized for the scenarios described above. Table 7.2 and figure 7.1 show the results for the different runs in RICE-99 and DICE-99 as well as the original DICE model. The definition of net economic impact used here is: The *net economic impact* of a policy is the sum across regions of the present value of consumption under that policy minus the present value of consumption in the base case. The present values are computed using the base case discount factors.

The RICE-99 model finds that the optimal policy produces a net economic gain of $198 billion. This is the present value of the gain to all regions, discounted back to 1995. Concentrating on the RICE-99 model results shown in table 7.2, we see that a delay of ten years leads to a trivially small net loss: $6 billion. This important result indicates that the loss from waiting and gathering more information is relatively small, assuming that action is appropriately taken in the future.

The next set of policies concerns emissions limitations. On theoretical grounds, one would expect these policies to be relatively inefficient because they target an inappropriate variable. Emissions are an inappropriate policy instrument because they are not of any intrinsic concern; they are intermediate variables connecting economic activity and the ultimate variable of concern, which is damages from climate

**Table 7.2**
Net economic impact (billions of 1990 U.S. dollars)

|  | RICE-99 | DICE-99 | Old Dice |
|---|---|---|---|
| *Base* | 0 | 0 | 0 |
| *Optimal* | | | |
| Policy in 1995 | 198 | 254 | 283 |
| Policy in 2005 | 192 | 246 | 254 |
| *Limit emissions* | | | |
| Global stabilization | −3,021 | −5,705 | −7,394 |
| Kyoto Protocol (a) | −120 | na | na |
| *Limit concentrations* | | | |
| Double CO$_2$ | −684 | −1,890 | na |
| *Limit temperature* | | | |
| 2.5 degree increase | −2,414 | −4,396 | na |
| 1.5 degree increase | −26,555 | −20,931 | −42,867 |
| *Geoengineering* (b) | 3,901 | 2,775 | 5,859 |

Notes: Source for Old Dice is Nordhaus 1994b, chapter 5, table 5.1.
(a) Annex I trading.
(b) Implemented by assuming that the damages from climate change are zero.

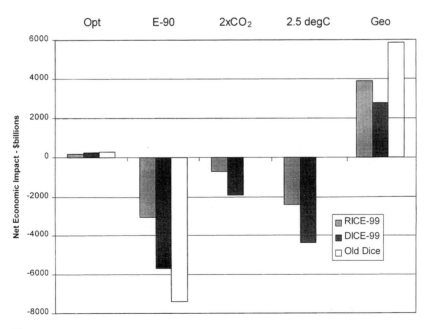

**Figure 7.1**
Global net economic impact

change. A policy that limits global emissions to 1990 levels has a discounted loss of $3 trillion in RICE-99. The Kyoto Protocol with Annex I trading has relatively small impact (a small loss) because RICE-99 projects relatively low emissions of Annex I regions; this also implies that the environmental gains from the Kyoto Protocol are small. (This point will be discussed below.) Emissions, concentrations, and temperature increases under the Kyoto Protocol are very close to the base policy because it has little impact on global emissions.

A policy of limiting $CO_2$ concentrations to double preindustrial levels has unfavorable net economic impacts, with a net loss of $0.7 trillion. A policy to limit temperature to 2.5° C is also quite costly. The present value of the net economic impact is about $2.5 trillion.

Finally, as is intuitively clear, geoengineering options that in effect remove atmospheric carbon at zero cost or neutralize the damages from climate change have highly positive net value. Estimates from RICE-99 indicate that the value is almost $4 trillion. This gives us a measure of the damages from climate change in the base policy.[6]

Table 7.3 shows the breakdown of costs, damages, and net benefits for the different policies in RICE-99. *Abatement costs* are defined as the difference between the present value of consumption in the base case and the present value of consumption under the policy assuming that the policy does not have any effect on the path of global mean temperature. The environmental benefits of the policy are then the sum of the abatement costs and the net economic impact. It is apparent that there are modest potential benefits from a successful climate change policy. The reduced damages from slowing climate change range from about $300 billion in present value in the optimal policy to $1.5 billion in the more ambitious emissions-limitation plan. Increases in production costs are also substantial, however. The optimal abatement policy incurs $98 billion in abatement costs, while inefficient plans such as

---

6. There are two potential ways of implementing a costless geoengineering policy in RICE-99. The first, which we have pursued, is to set the damage coefficients $\theta_{1,j}$ and $\theta_{2,j}$ in equation (2.16) to zero. This would correspond to a geoengineering policy that aims to exactly offset the increase in GHG concentrations. A second approach is to optimize the global mean temperature variable. Under this alternative, costless emissions (positive or negative) are used to achieve the temperature path that will give the highest discounted value of consumption. This second approach turns out to be difficult to implement in RICE-99, but by using DICE-99 we estimate that the optimal climate would have a present value of $1.7 trillion higher than the estimates presented here. Although aiming for the optimal climate is an intriguing approach, it requires a much deeper knowledge of impact than we possess, so we aim for the more modest goal of offsetting climate change.

**Table 7.3**
Abatement costs and environmental benefits of different policies (billions of 1990 U.S. dollars)

|  | Abatement cost | Environmental benefit | Net economic impact | Benefit/cost ratio |
|---|---|---|---|---|
| *Base* | 0 | 0 | 0 | na |
| *Optimal* | | | | |
| Policy in 1995 | −98 | 296 | 198 | 3.02 |
| Policy in 2005 | −92 | 283 | 192 | 3.08 |
| *Limit emissions* | | | | |
| Global stabilization | −4,533 | 1,512 | −3,021 | 0.33 |
| Kyoto Protocol | −217 | 96 | −120 | 0.44 |
| *Limit concentrations* | | | | |
| Double CO$_2$ | −1,365 | 681 | −684 | 0.50 |
| *Limit temperature* | | | | |
| 2.5 degree increase | −3,553 | 1,139 | −2,414 | 0.32 |
| 1.5 degree increase | −28,939 | 2,383 | −26,556 | 0.08 |
| *Geoengineering* | 0 | 3,901 | 3,901 | na |

stabilizing global emissions or limiting temperature increases to 2.5°C impose present value costs in the range of $3.5 to $4.5 trillion.

The benefit-cost ratios of different policies are shown in the last column of table 7.3. The optimal policy passes a cost-benefit test—it has a benefit-cost ratio of 3.0. The inefficient plans, by contrast, have benefit-cost ratios of 0.08 to 0.5. In judging these ratios, one must recall that we assume these policies are efficiently implemented. If inefficient implementation occurs (say through allocation of permits, exclusions, inefficient taxation, or regional exemptions), then the costs will rise and the benefit-cost ratio of even the optimal policy will quickly pass below 1.

Table 7.4 shows the regional breakdown of the net economic impact for different policies. In analyzing the regional impacts, we assume a revenue-neutral allocation of emissions permits or zero net permit revenue in each of the cases except the Kyoto Protocol, as explained in section two above. This assumption is important for the regional allocation of the costs and benefits of climate change policies.

To the extent that regions have relatively high emissions reductions under a policy or benefit from modest climate change (both of which

**Table 7.4**
Net economic impact of policies (difference from base, billions of 1990 U.S. dollars)

|  | Optimal | Limit to 1990 emissions | Limit to $2 \times CO_2$ concentrations | Limit temperature rise to 2.5 deg C | Geoengineering |
|---|---|---|---|---|---|
| USA | 22 | −946 | −305 | −885 | 82 |
| OHI | 26 | −139 | −6 | −131 | −391 |
| Europe | 126 | 258 | 162 | 121 | 1,943 |
| R&EE | −9 | −359 | −64 | −191 | −110 |
| MI | 19 | −300 | −103 | −304 | 620 |
| LMI | 5 | −512 | −122 | −341 | 549 |
| China | −10 | −425 | −74 | −226 | −21 |
| LI | 20 | −597 | −174 | −458 | 1,228 |
| Annex I | 164 | −1,187 | −212 | −1,085 | 1,524 |
| ROW | 34 | −1,834 | −472 | −1,329 | 2,377 |
| World | 198 | −3,021 | −684 | −2,414 | 3,901 |

occur in the case of the Russia and Eastern Europe region), the policy may lead to economic losses.

The major region to gain from climate policies is OECD Europe, which benefits from all policies, even the ones that have high global costs. In the optimal case, OECD Europe has over three-fifths of the net gain. These gains arise primarily because the region is highly sensitive to climate change, has a low discount rate, and pays little of the abatement costs under the policy of zero net permit revenue. Regions with high carbon intensities, high discount rates, and low vulnerability to global warming (such as Eastern Europe and China) have negative net impacts in the optimal case. Note that if there is trading of emissions permits, virtually any regional redistribution of the costs and benefits would be possible through the initial allocation of permits. This is not done in the current version for simplicity of presentation and calculation.

The final column in table 7.4 shows the net impacts of geoengineering; this is also approximately the net climate damages in the base case. The interesting result is that the major gains from geoengineering accrue to OECD Europe. As was shown in chapter 4, Russia, China, and Canada are likely to benefit from modest climate change of the kind found in the base case and have negative benefits from geoengineering.

The difference between the geoengineering results and the results for the other policies is so dramatic that it suggests that geoengineering should be more carefully analyzed. Table 7.3 indicates that a technological solution that would offset the climatic impacts of increasing greenhouse-gas concentrations would have a benefit of around $4 trillion in present value. In addition to its significant economic benefits, geoengineering also has important political advantages over the current approach of emissions reductions. Geoengineering does not require near-unanimous agreement among all major countries to have an effective policy; indeed, the United States could easily undertake geoengineering by itself if other countries would give their assent. Given its clear economic and political advantages, we believe that geoengineering should be much more carefully analyzed.

*Emissions Controls and Carbon Taxes*

Table 7.5 and figures 7.2 and 7.3 show the carbon taxes or permit prices in the different policies. In the runs analyzed here, the prices are internationally harmonized; this could occur in practice either through

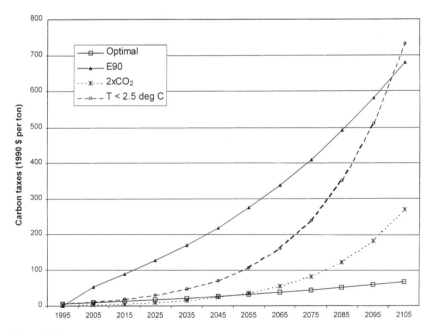

**Figure 7.2**
Carbon taxes: Alternative policies

**Table 7.5**
Carbon taxes in alternative policies (1990 U.S. dollars per ton carbon)

| | 1995 | 2005 | 2015 | 2025 | 2035 | 2045 | 2055 | 2065 | 2075 | 2085 | 2095 | 2105 |
|---|---|---|---|---|---|---|---|---|---|---|---|---|
| Optimal | 5.90 | 9.13 | 12.71 | 16.72 | 21.16 | 26.12 | 31.64 | 37.73 | 44.38 | 51.55 | 59.20 | 67.39 |
| Delayed optimal | 0.00 | 9.15 | 12.73 | 16.73 | 21.17 | 26.12 | 31.64 | 37.72 | 44.37 | 51.54 | 59.18 | 67.37 |
| Limit to 1990 emissions | 0.00 | 52.48 | 89.69 | 128.03 | 169.62 | 217.89 | 273.65 | 337.45 | 409.67 | 490.60 | 580.45 | 679.32 |
| Limit to 2 times $CO_2$ concentrations | 2.15 | 3.81 | 6.28 | 9.98 | 15.54 | 23.87 | 36.32 | 54.76 | 81.92 | 121.75 | 180.31 | 267.69 |
| Limit temperature | 6.73 | 11.79 | 19.20 | 30.27 | 46.71 | 71.19 | 107.44 | 160.74 | 238.35 | 350.28 | 509.65 | 732.03 |

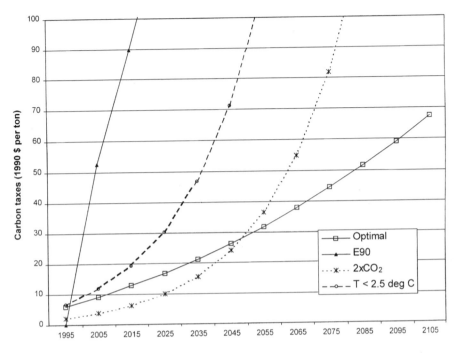

**Figure 7.3**
Carbon taxes: Alternative policies

harmonized taxes or a system of tradable emissions permits. The optimal policy has a carbon tax beginning at $6 per metric ton carbon for the first period (1990–99). The optimal tax rises in future years, reaching $13 per ton in 2015, $29 per ton in 2050, and $63 per ton carbon in 2100. For reference, a $10 per ton carbon tax will raise coal prices by $5.50 per short ton—about 30 percent of the current U.S. minemouth coal price. Further, a $10 per ton carbon tax would raise gasoline prices by about 2 U.S. cents per gallon.

The ten-year delay (not shown) has a zero carbon tax in the first period, but then is virtually indistinguishable from the optimal policy. The policy of no controls obviously has a zero carbon tax. The global emissions stabilization policy has steeply rising carbon taxes, hitting $200 per ton in the middle of the next century. Policies that stabilize $CO_2$ concentrations and temperature have initial carbon taxes close to those in the optimal policy, but they then rise sharply in the coming years. The optimal policy to meet these targets delays high carbon taxes to the future; reducing future emissions is a more cost-effective way to meet such targets both because it is less expensive in a present value

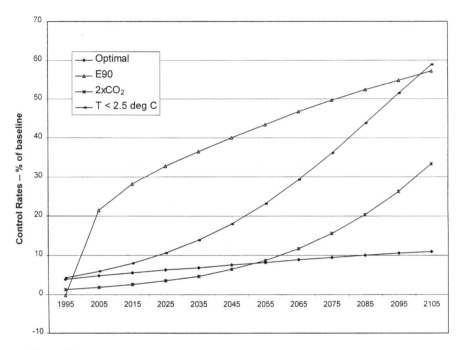

**Figure 7.4**
Emission control rates: Alternative policies

sense and because much of a current emission has been removed from the atmosphere when the target becomes a binding constraint.

Table 7.6 and figure 7.4 show the control rate for $CO_2$ for the different policies. They show the extent to which GHG emissions are reduced below their reference levels. In the optimal policy, the rate of emissions reduction begins at a low rate of about 4 percent of emissions and climbs slowly over the next century, reaching about 11 percent of baseline emissions by 2100. The three environmental policies (limiting emissions, concentrations, and temperature) start with relatively low emissions controls but then climb sharply to emissions-control rates between 33 and 59 percent by the end of the next century.

Figure 7.5 shows the regional control rates in the optimal policy. One interesting feature is that the control rates fall into two general groups—those for high income regions and those for low-income economies and economies in transition. The control rates for the latter regions are generally more than twice those of the high-income regions. The reason for the difference is that energy is generally much more highly taxed in high-income countries, while it is often subsidized in

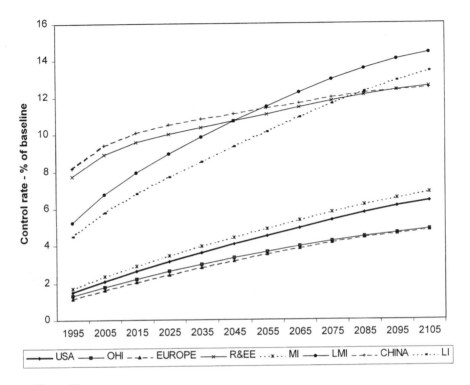

**Figure 7.5**
Optimal emissions control rate by region

low-income countries, so that it is less expensive to reduce emissions
in the low-taxed regions. It is interesting to note that this pattern is
exactly the *opposite* of the prescription in the Kyoto Protocol, which has
the initial emissions reductions in the high-income Annex I countries.

### Emissions, Concentrations, and Climate Change

We next show some of the climatic details of the model runs. Table 7.7
and figure 7.6 show the aggregate industrial $CO_2$ emissions, while
figure 7.7 shows the reference emissions for different regions. Baseline
industrial $CO_2$ emissions in RICE-99 project steady growth to about 13
GtC annually in 2100. In the optimal policy, emissions are limited to
11.5 GtC in 2100. By comparison, emissions are only 9 GtC in 2100 for
limiting concentrations, and 5.8 GtC in 2100 for limiting temperature to
2.5°C.

Atmospheric concentrations of $CO_2$ are shown in figure 7.8. Beginning at an atmospheric concentration of 735 GtC (345 ppm) in 1995,

**Table 7.6**
Emissions control rates in alternative policies (reduction in emissions as percent of baseline emissions)

|  | 1995 | 2005 | 2015 | 2025 | 2035 | 2045 | 2055 | 2065 | 2075 | 2085 | 2095 | 2105 |
|---|---|---|---|---|---|---|---|---|---|---|---|---|
| Optimal | 3.9 | 4.8 | 5.6 | 6.2 | 6.9 | 7.5 | 8.2 | 8.8 | 9.4 | 10.0 | 10.5 | 10.9 |
| Limit to 1990 emissions | 0.0 | 21.6 | 28.2 | 32.8 | 36.5 | 40.1 | 43.4 | 46.6 | 49.6 | 52.3 | 54.9 | 57.2 |
| Limit to 2 times $CO_2$ concentrations | 1.3 | 1.9 | 2.5 | 3.5 | 4.7 | 6.4 | 8.7 | 11.7 | 15.5 | 20.4 | 26.3 | 33.3 |
| Limit temperature rise to 2.5 deg C | 4.2 | 6.0 | 8.0 | 10.5 | 13.8 | 18.0 | 23.1 | 29.2 | 36.1 | 43.7 | 51.5 | 59.0 |

**Table 7.7**
Industrial $CO_2$ emissions in alternative policies (GtC per year)

|  | 1995 | 2005 | 2015 | 2025 | 2035 | 2045 | 2055 | 2065 | 2075 | 2085 | 2095 | 2105 |
|---|---|---|---|---|---|---|---|---|---|---|---|---|
| Base | 6.2 | 7.2 | 7.9 | 8.4 | 8.9 | 9.5 | 10.0 | 10.6 | 11.2 | 11.9 | 12.6 | 13.3 |
| Optimal | 5.9 | 6.9 | 7.5 | 7.9 | 8.3 | 8.7 | 9.2 | 9.7 | 10.2 | 10.7 | 11.3 | 11.8 |
| Limit to 1990 emissions | 6.2 | 5.7 | 5.7 | 5.7 | 5.7 | 5.7 | 5.7 | 5.7 | 5.7 | 5.7 | 5.7 | 5.7 |
| Limit to 2 times $CO_2$ concentrations | 6.1 | 7.1 | 7.7 | 8.1 | 8.5 | 8.9 | 9.1 | 9.4 | 9.5 | 9.5 | 9.3 | 8.8 |
| Limit temperature rise to 2.5 deg C | 5.9 | 6.8 | 7.3 | 7.6 | 7.7 | 7.8 | 7.7 | 7.5 | 7.2 | 6.7 | 6.1 | 5.4 |

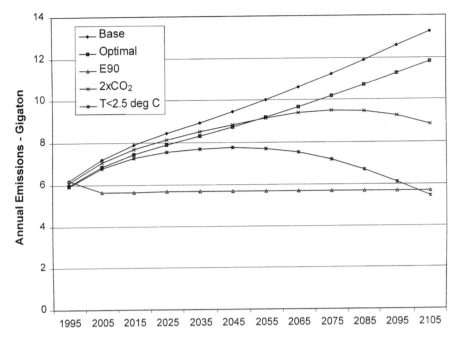

**Figure 7.6**
Industrial $CO_2$ emissions: Alternative policies

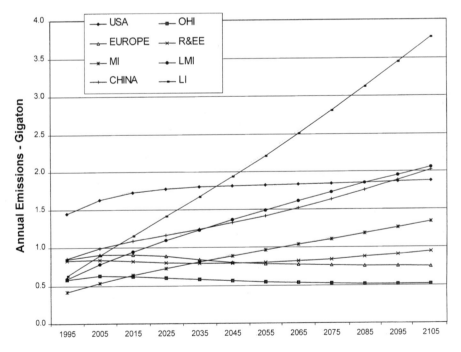

**Figure 7.7**
Regional industrial $CO_2$ emissions in base case

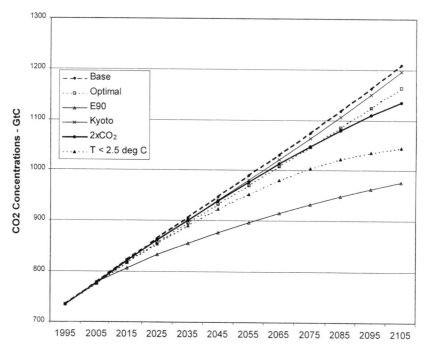

**Figure 7.8**
$CO_2$ concentrations: Alternative policies

baseline concentrations rise to 1187 GtC (557 ppm) in 2100. In the
optimal control case, concentrations are limited to 1145 GtC (538 ppm)
in 2100. The Kyoto concentration shown in figure 7.8 and discussed in
the next chapter is very close to the base case.

It is interesting to note that the emissions and concentrations projections in RICE-99 are well below those in many current projections. For
comparison, of the ninety scenarios examined in IPCC 1995, the median
projection for 2100 was around 22 GtC.[7] In the often-cited IPCC IS92a
scenario, the 2100 carbon dioxide concentration is 710 ppmv (1,500 GtC).
On the other hand, many of these scenarios were prepared before the
breakup of the Soviet Union and contained high rates of economic
growth and low rates of decarbonization. It will be necessary to wait for
the next generation of studies to determine whether the relatively low
emissions projections in RICE-99 are an aberration or a harbinger.

The temperature trend in the base RICE-99 run, however, is close
to the IPCC projections developed in the early 1990s. The baseline

7. See IPCC 1995, figure 6.2.

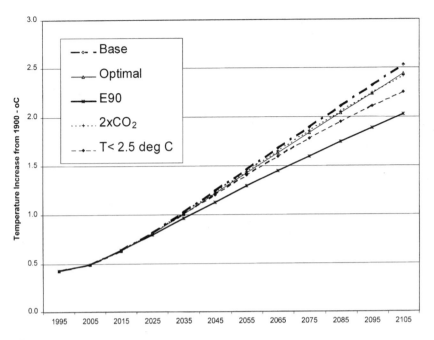

**Figure 7.9**
Global mean temperature

temperature increase relative to 1900 in RICE-99 is 0.43°C in 1995 and rises to 2.42°C by 2100, for an increase of 2.0°C. This increase compares with a baseline warming used in IPCC [1996a] of 2.0°C in 2100 relative to 1990 with a climate sensitivity of 2.5°C. RICE-99 projects higher global temperatures despite lower projected emissions than the IPCC because RICE-99 uses a higher equilibrium temperature sensitivity of 2.9°C for a doubling of $CO_2$.

The effect of alternative policies on the projected global mean temperature is shown in table 7.8 and figure 7.9. All runs have very similar temperature trajectories through the middle of the next century. After 2050, the emissions-limitation scenario begins to head down relative to the other paths. One of the surprising results of virtually all policies is how little they affect the temperature trajectory in the next century. The optimal policy reduces global mean temperature by 0.09°C relative to the baseline in 2100; however, the policies have a more substantial impact in the next century. The temperature reduction in 2200 relative to the baseline of the optimal, concentration target, and temperature target are 0.20, 0.88, and 1.37°C, respectively.

**Table 7.8**
Temperature in alternative policies (difference in global mean atmospheric temperature from 1900, degrees C)

|  | 1995 | 2005 | 2015 | 2025 | 2035 | 2045 | 2055 | 2065 | 2075 | 2085 | 2095 | 2105 |
|---|---|---|---|---|---|---|---|---|---|---|---|---|
| Base | 0.43 | 0.49 | 0.63 | 0.82 | 1.03 | 1.25 | 1.46 | 1.68 | 1.89 | 2.11 | 2.32 | 2.53 |
| Optimal | 0.43 | 0.49 | 0.63 | 0.81 | 1.01 | 1.22 | 1.43 | 1.63 | 1.84 | 2.04 | 2.24 | 2.44 |
| Limit to 1990 emissions | 0.43 | 0.49 | 0.63 | 0.80 | 0.96 | 1.13 | 1.29 | 1.45 | 1.60 | 1.75 | 1.89 | 2.02 |
| Limit to 2 times $CO_2$ concentrations | 0.43 | 0.49 | 0.63 | 0.82 | 1.02 | 1.23 | 1.45 | 1.65 | 1.85 | 2.05 | 2.24 | 2.42 |
| Limit temperature rise to 2.5 deg C | 0.43 | 0.49 | 0.63 | 0.81 | 1.01 | 1.21 | 1.41 | 1.60 | 1.78 | 1.95 | 2.11 | 2.25 |

One puzzling feature of these results is the modest impact that the optimal policy or even more ambitious policies make upon the concentration and temperature trajectories. The first reason for the modest effect is straightforward. According to our estimates, the impact of warming upon the global economy is relatively small, amounting to around 2 percent of global output for a 2.5° C average warming. By contrast, the abatement costs of significant reductions in GHGs are high. The interaction of small benefits and large costs is that an optimal policy has little effect on the near-term temperature increase.

Two other factors lead to the small decrease in the extent of warming in the optimal policy. First, there is a great deal of momentum of climate change given the existing degree of buildup of GHGs and the lags in the response of the climate to GHG increases. For example, consider the policy of stabilizing global emissions at 1990 rates—an extremely ambitious target that requires reducing $CO_2$ emissions by over 40 percent below the baseline by the middle of the next century and costing $4.5 trillion in discounted abatement expenditures. Even with all this cost, global temperatures would still rise by slightly more than 2° C above 1900 by 2100.

Second, the relationship between GHG concentrations and warming is nonlinear. According to scientific studies, the relationship between equilibrium warming and $CO_2$ concentrations is approximately logarithmic. This implies that moving from 300 to 315 ppm of $CO_2$ increases equilibrium temperature by 0.205° C while moving from 600 to 615 ppm of $CO_2$ increases equilibrium temperature by only 0.104° C. The implication of this nonlinear relationship is that policies that produce a small decrease in $CO_2$ concentrations have a relatively small impact upon the path of temperature. This result is the opposite of the usual diminishing returns seen almost everywhere in economic systems.

*Other Economic Variables*

The model has a wide variety of other projections that are necessary for a full analysis. They include physical output (such as $CO_2$ emissions) as well as economic values (such as the values of output and consumption). Figure 7.10 shows per capita output for the eight regions, while the trends in regional carbon intensities are shown in figure 7.11. Regional emissions are also shown in figure 7.7.

Two points about the trends should be noted. First, the model assumes continued rapid economic growth in the years ahead. This

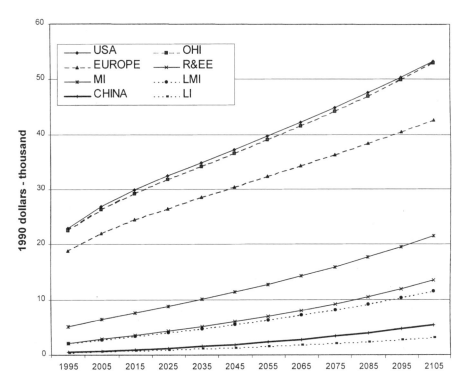

**Figure 7.10**
Per capita income in base run

growth will lead to increased emissions, but it will also improve living standards and provide resources for coping with greenhouse warming. A second important feature of the RICE-99 model projections is that $CO_2$ emissions in the high-income regions are projected to be relatively flat over the next century. This is the result of continuing decarbonization plus slower population growth in high-income countries. This trend has important implications for the Kyoto Protocol, because it constrains only high-income countries.

*Comparison of DICE-99 and DICE-94*

Finally, the extent to which projections and policies differ among the current RICE/DICE-99 models is examined and compared those with the original DICE model (Nordhaus 1994b). Although DICE-99 tracks RICE-99 closely as far as important variables are concerned over the next century (see tables 5.1 and 5.2). Table 7.2 shows that in some

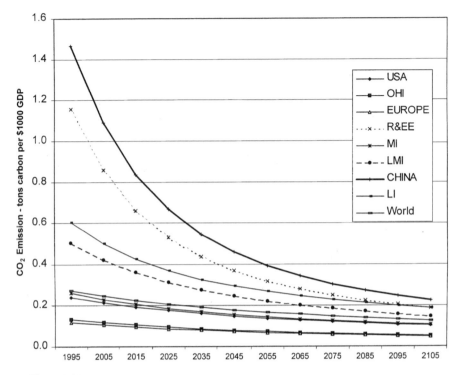

**Figure 7.11**
Industrial carbon intensity: Base case

cases the net economic impact of a policy differs substantially between the two models. The policy ranking that comes from the two models, however, is the same.

Another interesting comparison is between the original and current versions of DICE. The net economic impact of policies has changed surprisingly little across model versions, except for cases that move far away from the baseline.

The carbon taxes for the first period are quite close in all three models, with the optimal carbon tax being $5.53 per ton in the original DICE model and $5.90 in RICE-99 and DICE-99. The optimal carbon tax turns out to be one of the most robust numbers through the evolving family of DICE and RICE models.

# 8          Economic Analysis of the Kyoto Protocol

The last chapter reviewed a wide range of alternative approaches for dealing with the threat of climate change. This chapter focuses on the current approach that has been adopted under international agreements, namely the Kyoto Protocol. The discussion begins with a description of the protocol and then analyzes its impact on economic and environmental outcomes.[1]

## Climate-Change Policy and the Kyoto Protocol

Governments have struggled to find policies that can at the same time satisfy the demands of electoral politics and meet the needs for responsible global stewardship. The initial response of nations to the threat of global warming was the Framework Convention on Climate Change (FCCC), which was issued at the Rio Summit of 1992. Under the FCCC, Annex I countries (high-income nations plus the former Soviet Union and Eastern European countries) committed on a voluntary basis to limit their concentrations of GHGs to 1990 levels. The FCCC left open almost all the important questions, such as the environmental, economic, and political components of such a commitment.[2]

It soon became apparent that the voluntary approach under the FCCC was producing next to nothing in real policy measures. Moreover, some countries, particularly the United States, were experiencing rapid growth in $CO_2$ emissions. This led the advocates of strong policy measures to pursue binding commitments, which led to the Kyoto Protocol of December 1997. The key provision of the Kyoto Protocol is

---

1. An earlier version of this chapter, using the RICE-98 model, appeared in Nordhaus and Boyer 1999.
2. A full discussion of the FCCC can be found at the website http:// www.unfccc.de/. The text and discussion of the Kyoto Protocol can also be found at that site.

article 3, which states that Annex I countries will on average reduce their emissions of GHGs by 5 percent relative to 1990 levels by the budget period 2008–2012, with each country facing an individual emissions limit (see chapter 7, the second section, the fourth subsection).[3]

Both economic theory and historical experience have shown that allowing economic agents to trade—in this case, to trade national emissions-reduction permits—can substantially reduce the cost of meeting an aggregate quantitative reduction target. The United States therefore proposed international emissions trading. The trading provision is contained in article 6, which reads: "For the purpose of meeting its commitments under article 3, any party included in Annex I may transfer to, or acquire from, any other such party emission reduction units . . . provided that: . . . the acquisition of emission reduction units shall be supplemental to domestic actions for the purposes of meeting commitments under article 3." This provision gives Annex I nations the right to trade emissions units. It is haunted by the vague and troubling provision, however, that the acquired permits will be supplemental to domestic actions. In other words, nations can buy only part of their emissions reductions, although the allowable amounts are unspecified in the protocol.

An additional provision introduces the possibility of offsets from developing countries. Article 12 defines a *clean development mechanism*, under which "(a) parties not included in Annex I will benefit from project activities resulting in certified emission reductions; and (b) parties included in Annex I may use the certified emission reductions accruing from such project activities to contribute to compliance with part of their quantified emission limitation and reduction commitments . . . Emission reductions resulting from each project activity shall be certified . . . on the basis of . . . real, measurable, and long-term benefits related to the mitigation of climate change [and] reductions in emissions that are additional to any that would occur in the absence of the certified project activity."[4] Some have interpreted this as a green light to include trading with developing countries, but the need to ensure additionality and to certify each transaction probably means it will lead to only a small fraction of potential trades.

---

3. As discussed below, the protocol opens the door for possible emissions trading and other cooperative schemes, so it might be possible for a country to meet its emissions limit even if its actual emissions exceed that limit.

4. All citations of the protocol have omitted provisions that are not relevant to the present analysis, such as the need for consent and the monitoring by international bodies.

A further complication involves GHG emissions other than those from energy use. The Kyoto Protocol has provisions for five other gases as well as for the potential for enhancing sinks. Specialists are working to understand the potential offsets that might come from these additional actions and to clarify the treatment of carbon sinks in the treaty.

The various versions of the Kyoto Protocol considered below make a variety of assumptions about Annex I regions' abilities to reduce their own abatement costs by buying emissions reductions elsewhere. We assume that other gases and sinks are neutral. More precisely, we assume that reductions in other GHGs and increased sequestration in carbon sinks for each region are such that the percentage reduction in industrial emissions required to meet Kyoto is below 1990 industrial emissions by the same fraction as the overall GHG target is below the 1990 level.

## Economic Analysis of the Kyoto Protocol

Earlier chapters discussed the details of the RICE-99 model. We now discuss the modifications of RICE-99 needed to analyze the Kyoto Protocol. We analyze in this chapter a number of different approaches to implementing the Kyoto Protocol and compare the different approaches with the optimal run described in the last chapter. Table 8.1 shows the major runs analyzed in this chapter. Most of them require no discussion, but a few details need elaboration.

**Table 8.1**
Runs for analysis of Kyoto Protocol

---

1. *Reference:* no controls

2. *Optimal:* sets emissions by region and period to balance the costs and benefits of emissions reductions

3. *Kyoto emissions limitations:*
    a. No trade: no trade among 4 major Annex I blocs
    b. OECD trade: emissions trading limited to OECD countries
    c. Annex I trade: emissions trading limited to Annex I countries
    d. Global trade: emissions trade among all regions

4. *Cost effectiveness benchmarks:*
    a. Limit atmospheric concentrations to those resulting from the Kyoto Protocol case 3c (for the period after 2050)
    b. Limit global mean temperature to that resulting from the Kyoto Protocol case 3c (for the period after 2100)

---

Note: This list shows the runs examined in the analysis of the Kyoto Protocol.

The reference case (policy 1) and the optimal case (policy 2) were described in chapter 7. The cases denoted Kyoto emissions limitations take the emissions permit allocations agreed upon in the 1997 Kyoto Protocol and extend them indefinitely for Annex I regions—this might be called "Kyoto forever."[5] Four variants of the Kyoto Protocol are analyzed here. Under the "no-trade" run 3a, no trading of emissions permits is allowed among the four Annex I regions included in RICE-99, and there are no offsets with the non-Annex I regions. Under the "OECD-trade" runs (run 3b), emissions trading is allowed only among the OECD regions.[6] The "Annex I-trade" case (run 3c) allows trading among all Annex I regions. This is the same as policy 4 in the previous chapter. The global trading policy (run 3d) extends the umbrella of trading to all regions. In this case, the non-Annex I regions receive emissions permits equal to their baseline emissions from run 1, but regions are then allowed to sell any emissions rights that exceed their actual emissions. Each of these runs has serious implementation issues, but these are ignored in this analysis.

Although the Kyoto emissions limitation cases are referred to as types of trading regimes, each of these could be implemented as a fiscal regime in which carbon taxes are made uniform across a trading bloc and the tax revenues shared across regions in a bloc.

It should be emphasized that global trading in case 3d is a radical extension of the Kyoto Protocol and contains crucial and problematical assumptions about the behavior of non-Annex I countries. In principle, each non-Annex-I country will be better off by agreeing to this limit-and-trade procedure; it can do no worse than simply consuming its permits and can do better by reducing its low marginal-cost sources and selling the permits at the world price. This assumption is questionable in practice, however, for three reasons: (1) the difficulty of estimating and assigning the appropriate baseline emissions, (2) the need to ensure compliance among countries with weak governance structures, and (3) the potential for countries to repudiate their commitments in the future.

In addition to the Kyoto runs, two alternative cases are presented that are useful for assessing the efficiency of different approaches. The

---

5. See chapter 7, section two, part four for description of Annex I in RICE-99.

6. Strictly speaking, the "OECD" here consists of the US, OHI, and OECD Europe regions, the high-income regions. This includes the actual OECD less Mexico, South Korea, Poland, Hungary, and the Czech Republic plus Singapore, Israel, Hong Kong, and a handful of small island nations.

emissions objectives of the Kyoto Protocol are not based on any ultimate environmental objective; instead, they are the simple and easily understood guidelines of holding emissions constant. The emissions objectives can be translated into more meaningful environmental objectives by examining the consequences of the Kyoto Protocol for $CO_2$ concentrations and for global temperature. Run 4a finds a Pareto optimal carbon tax trajectory subject to the concentrations target implicit in the Kyoto emissions limitations; in this policy, concentrations are constrained to be at the same level implied by the Kyoto Protocol after 2050. Run 4b takes the same approach for global temperature, where temperature is constrained to be at the same level as that implicit in the Kyoto Protocol after 2100. (These dates are selected to take account of the lags between emissions and the two other objectives.) Runs 4a and 4b allow one to ask how cost-effective the different approaches are in attaining the environmental objectives embodied in the Kyoto Protocol. In both runs 4a and 4b, the entire world is treated as one trading bloc, and net permit revenue is held to zero by assigning to each region emissions permits equal to its emissions.

As one moves from the no-trade case to the global-trading case, the where-efficiency of the Kyoto Protocol is improved. Each step from global trading to case 4a and from 4a to 4b improves on when-efficiency. But even case 4b is not why-efficient, as a comparison with the optimal run will demonstrate, since the temperature target chosen has no grounding in optimization (see discussion of types of efficiency in chapter 7, section one).

**Major Results**

This section presents the results of the analysis of the Kyoto Protocol using RICE-99. Important conclusions are highlighted in this summary: the ranking of policies, the optimal carbon price, and a revised view of the climate-change problem.

*Environmental Variables*

The first set of results pertains to the major environmental variables: emissions, concentrations, and global temperature increases. Figure 8.1 and table 8.2 show global industrial $CO_2$ emissions for the major cases. The overall level of abatement in the early years under the Kyoto Protocol is close to that of the optimal program; in 2015 global emissions

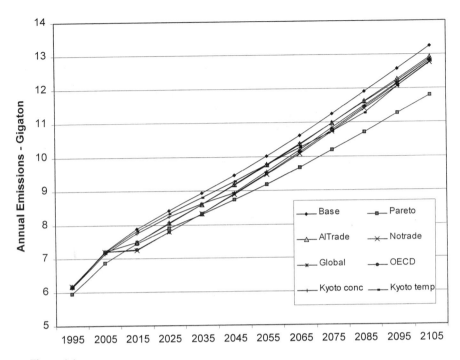

**Figure 8.1**
Global industrial $CO_2$ emissions

under the Annex I variant are 7.50 GtC compared to 7.45 GtC in the optimal case. The rate of emissions reduction falls over time, however, because of the exclusion of low-income countries from the Kyoto Protocol. By 2105, under the Kyoto Protocol, emissions are higher than under the optimal program. The fact that the initial level of global emissions reductions is close to the optimal in RICE-99 implies that the Kyoto Protocol with global trading would begin relatively efficiently, while other forms of implementation will prove inefficient because of where-inefficiency in allocation across different countries (particularly the omission of developing countries).

The buildup of $CO_2$ concentrations projected in different scenarios is shown in figure 8.2. In the base run of RICE-99, concentrations in 2100 are 557 ppm (1187 GtC). In the optimal run, concentrations in 2100 are 538 ppm (1145 GtC), while those in the Annex I version of the Kyoto Protocol are 551 ppm (1174 GtC). The Kyoto Protocol has higher concentrations than the optimal run because it leaves uncontrolled the emissions in the low-income regions. The Kyoto Protocol makes little headway in reducing $CO_2$ concentrations under any trading scheme.

**Table 8.2**
Industrial carbon emissions in Kyoto runs

| | Industrial carbon emissions, 2015 (GtC) | | | | | | | |
|---|---|---|---|---|---|---|---|---|
| | Base | Optimal | Kyoto conc. | Kyoto temp. | Global | AI trade | OECD | No trade |
| USA | 1.73 | 1.68 | 1.72 | 1.72 | 1.69 | 1.61 | 1.36 | 1.29 |
| OHI | 0.62 | 0.61 | 0.62 | 0.62 | 0.61 | 0.59 | 0.51 | 0.52 |
| Europe | 0.91 | 0.89 | 0.90 | 0.90 | 0.89 | 0.85 | 0.75 | 0.81 |
| R&EE | 0.82 | 0.74 | 0.80 | 0.81 | 0.75 | 0.63 | 0.82 | 0.82 |
| MI | 0.64 | 0.62 | 0.63 | 0.63 | 0.62 | 0.64 | 0.64 | 0.64 |
| LMI | 0.94 | 0.87 | 0.92 | 0.93 | 0.88 | 0.95 | 0.95 | 0.95 |
| China | 1.08 | 0.97 | 1.05 | 1.07 | 0.98 | 1.08 | 1.08 | 1.08 |
| LI | 1.15 | 1.08 | 1.13 | 1.14 | 1.08 | 1.15 | 1.15 | 1.15 |
| Annex I | 4.07 | 3.92 | 4.03 | 4.05 | 3.93 | 3.68 | 3.44 | 3.44 |
| ROW | 3.82 | 3.54 | 3.74 | 3.78 | 3.57 | 3.82 | 3.82 | 3.82 |
| World | 7.89 | 7.45 | 7.77 | 7.83 | 7.50 | 7.50 | 7.26 | 7.26 |

| | Industrial carbon emissions, 2105 (GtC) | | | | | | | |
|---|---|---|---|---|---|---|---|---|
| | Base | Optimal | Kyoto conc. | Kyoto temp. | Global | AI trade | OECD | No trade |
| USA | 1.87 | 1.75 | 1.83 | 1.83 | 1.84 | 1.71 | 1.54 | 1.29 |
| OHI | 0.51 | 0.49 | 0.50 | 0.50 | 0.50 | 0.48 | 0.44 | 0.51 |
| Europe | 0.75 | 0.71 | 0.74 | 0.73 | 0.74 | 0.70 | 0.64 | 0.75 |
| R&EE | 0.94 | 0.83 | 0.91 | 0.90 | 0.91 | 0.79 | 0.95 | 0.95 |
| MI | 1.33 | 1.24 | 1.30 | 1.30 | 1.31 | 1.34 | 1.34 | 1.34 |
| LMI | 2.05 | 1.76 | 1.97 | 1.96 | 1.97 | 2.07 | 2.08 | 2.08 |
| China | 2.02 | 1.77 | 1.94 | 1.94 | 1.95 | 2.03 | 2.04 | 2.04 |
| LI | 3.77 | 3.26 | 3.62 | 3.60 | 3.63 | 3.80 | 3.80 | 3.80 |
| Annex I | 4.07 | 3.77 | 3.98 | 3.98 | 4.00 | 3.68 | 3.57 | 3.50 |
| ROW | 9.17 | 8.03 | 8.82 | 8.80 | 8.86 | 9.23 | 9.25 | 9.26 |
| World | 13.25 | 11.80 | 12.80 | 12.77 | 12.86 | 12.91 | 12.83 | 12.76 |

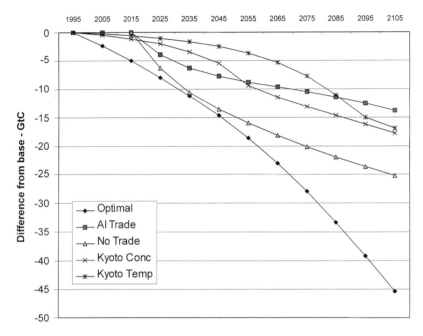

**Figure 8.2**
Atmospheric CO$_2$ concentration

The impact of the Kyoto Protocol on global temperature is exceed-
ing modest, especially for the first century (see figure 8.3). Table 8.3
shows that the reduction in global mean temperature in the Annex I
case is 0.03°C relative to the base case in 2100. This compares with a
difference of 0.17°C from the Kyoto Protocol calculated by Wigley 1998,
which assumes the IPCC IS92a emissions scenario as a reference. The
temperature reduction is so small in RICE-99 because of the flat base-
case trajectory of emissions (and consequently the small impact on
emissions) of Annex I regions projected in the current study compared
to the IPCC emissions scenario.

To summarize the key results:

1. Emissions reductions under the Kyoto Protocol are extremely
modest. Indeed, they are less than those projected under the optimal
program. The reason is primarily that emissions in the high-income
countries are projected to grow quite slowly, whereas the more rapidly
growing emissions in developing countries are uncontrolled under the
Kyoto Protocol.

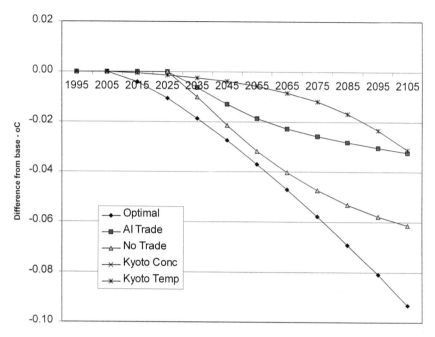

**Figure 8.3**
Global temperature increase

**Table 8.3**
Comparison of global mean temperature increase in different runs

|  | Atmospheric temperature (degrees C increase from 1900) | | |
|---|---|---|---|
|  | 1995 | 2055 | 2105 |
| Base | 0.43 | 1.46 | 2.53 |
| Optimal | 0.43 | 1.43 | 2.44 |
| Kyoto concentrations | 0.43 | 1.45 | 2.49 |
| Kyoto temperature | 0.43 | 1.46 | 2.50 |
| Global | 0.43 | 1.45 | 2.50 |
| AI trade | 0.43 | 1.45 | 2.50 |
| OECD | 0.43 | 1.43 | 2.47 |
| No trade | 0.43 | 1.43 | 2.47 |

2. The Kyoto Protocol is projected to have a very modest impact on $CO_2$ concentrations and on global warming. Because the Kyoto Protocol policy is not designed to cap the emissions of non-Annex I countries, the long-run impact of the Kyoto Protocol on carbon emissions and global temperature is extremely small.

3. In the short run, the global emissions, concentrations, and warming under the Kyoto Protocol are close to those in the optimal policy.

*Economic Variables*

**Carbon taxes.** One of the most useful measures of the strictness of climate-change policy is the carbon tax that would be generated by the policy. Table 8.4 and figure 8.4 show the carbon taxes for the major cases. The optimal and global-trading cases have relatively low carbon taxes. The carbon tax in the global-trading case is $11 per ton in 2015,

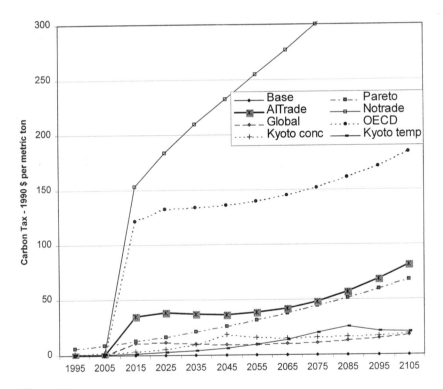

**Figure 8.4**
Carbon taxes in different policies

**Table 8.4**
Comparison of carbon taxes, 2015 and 2105, in different approaches to Kyoto Protocol

| | Base | Optimal | Kyoto concentrations | Kyoto temperature | Global trade | Annex I trade | OECD trade | No trade |
|---|---|---|---|---|---|---|---|---|
| *Carbon tax, 2015 (1990 U.S. dollars per metric ton carbon)* | | | | | | | | |
| USA | 0.00 | 12.71 | 4.99 | 1.65 | 11.17 | 34.52 | 121.31 | 152.75 |
| OHI | 0.00 | 12.71 | 4.99 | 1.65 | 11.17 | 34.52 | 121.31 | 108.42 |
| Europe | 0.00 | 12.71 | 4.99 | 1.65 | 11.17 | 34.52 | 121.31 | 69.31 |
| R&EE | 0.00 | 12.71 | 4.99 | 1.65 | 11.17 | 34.52 | 0.00 | 0.00 |
| MI | 0.00 | 12.71 | 4.99 | 1.65 | 11.17 | 0.00 | 0.00 | 0.00 |
| LMI | 0.00 | 12.71 | 4.99 | 1.65 | 11.17 | 0.00 | 0.00 | 0.00 |
| China | 0.00 | 12.71 | 4.99 | 1.65 | 11.17 | 0.00 | 0.00 | 0.00 |
| LI | 0.00 | 12.71 | 4.99 | 1.65 | 11.17 | 0.00 | 0.00 | 0.00 |
| Annex I | 0.00 | 12.71 | 4.99 | 1.65 | 11.17 | 34.52 | — | — |
| ROW | 0.00 | 12.71 | 4.99 | 1.65 | 11.17 | 0.00 | 0.00 | 0.00 |
| World | 0.00 | 12.71 | 4.99 | 1.65 | 11.17 | 34.52 | 121.31 | 152.75 |

| | Base | Optimal | Kyoto concentrations | Kyoto temperature | Global trade | Annex I trade | OECD trade | No trade |
|---|---|---|---|---|---|---|---|---|
| *Carbon tax, 2105 (1990 U.S. dollars per metric ton carbon)* | | | | | | | | |
| USA | 0.00 | 67.39 | 24.57 | 20.53 | 17.63 | 81.61 | 184.33 | 378.01 |
| OHI | 0.00 | 67.39 | 24.57 | 20.53 | 17.63 | 81.61 | 184.33 | 0.00 |
| Europe | 0.00 | 67.39 | 24.57 | 20.53 | 17.63 | 81.61 | 184.33 | 0.00 |
| R&EE | 0.00 | 67.39 | 24.57 | 20.53 | 17.63 | 81.61 | 0.00 | 0.00 |
| MI | 0.00 | 67.39 | 24.57 | 20.53 | 17.63 | 0.00 | 0.00 | 0.00 |
| LMI | 0.00 | 67.39 | 24.57 | 20.53 | 17.63 | 0.00 | 0.00 | 0.00 |
| China | 0.00 | 67.39 | 24.57 | 20.53 | 17.63 | 0.00 | 0.00 | 0.00 |
| LI | 0.00 | 67.39 | 24.57 | 20.53 | 17.63 | 0.00 | 0.00 | 0.00 |
| Annex I | 0.00 | 67.39 | 24.57 | 20.53 | 17.63 | 81.61 | — | — |
| ROW | 0.00 | 67.39 | 24.57 | 20.53 | 17.63 | 0.00 | 0.00 | 0.00 |
| World | 0.00 | 67.39 | 24.57 | 20.53 | 17.63 | 81.61 | 184.33 | 378.01 |

then rises to about $16 per ton in 2100; this compares with our estimate of the optimal carbon tax, which is $13 per ton in 2015 and rises to $63 per ton in 2100. The prices indicate that the Kyoto Protocol starts out with an efficient *global* rate of emissions reduction but then fails to make sufficient further reductions in the coming decades.

Policies that restrict trade have higher average carbon taxes, with the variation of prices among regions leading to an inefficient distribution of abatement. The Annex I trading case has sharply rising carbon taxes in the Annex I regions, starting at $35 per ton in 2015 and rising to around $82 per ton at the end of the next century. These prices hold for Annex I regions; carbon taxes in non-Annex I regions are zero.

Russia and Eastern Europe play a crucial role in the Annex I version of the Kyoto Protocol. Baseline emissions in these countries are below their Kyoto Protocol limits; this provides an enormous pool of potential emissions reductions that keeps carbon taxes in the OECD region down in the Annex I case. As shown in figure 8.4, the carbon taxes for the no-trade version of the Kyoto Protocol are significantly higher than the Annex I case. For example, the U.S. 2015 carbon tax is $153 per ton for the no-trading case, rising to about $360 per ton by 2100. These numbers are so large that they cast a fairy-tale (or perhaps horror-story) quality to the analysis. For example, by the middle of the next century, annual U.S. carbon tax revenues are around $300 billion dollars in the no-trade version. In the Annex I case, the United States is transferring about $20 billion annually to other regions through purchases of carbon emissions permits in the next century.

**Overall abatement costs.** The next set of issues concerns the economic impact of alternative policies. The present value of total abatement costs is shown in figure 8.5 and table 8.5. The present value of abatement (which excludes damages) ranges from a low of $59 billion in the global-trading case, to $217 billion in the Annex I-trading case, to a high of $884 billion in the no-trade case. Clearly, there are enormous stakes involved in policies to control global warming. (See chapter 7, section three, the first subsection for definition of abatement costs.)

It is interesting to compare the costs of different regimes with the minimum global cost of meeting the Kyoto temperature trajectory. We estimate that the trajectory can be attained at a minimum cost of $5

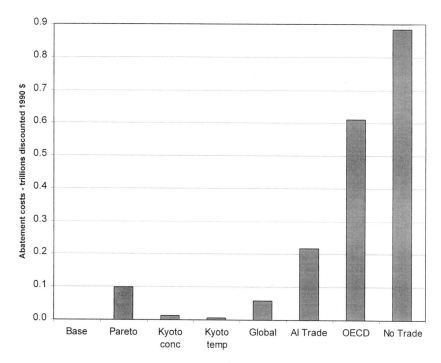

**Figure 8.5**
Abatement costs in different strategies

**Table 8.5**
Discounted abatement costs in different strategies

| Strategy | Discounted costs (billions of 1990 U.S. dollars) |
|---|---|
| Base | 0 |
| Optimal | 98 |
| Kyoto Protocol | |
|   Global | 59 |
|   AI trade | 217 |
|   OECD | 611 |
|   No trade | 884 |
| Kyoto concentration limit | 12 |
| Kyoto temperature limit | 5 |

Note: Table 8.5 shows the discounted global costs of different targets or control strategies. The estimates are the discounted consumption in the base case minus the discounted consumption in the case in question, where these have been calculated excluding the environmental benefits of controls.

billion. The global trading scenario is relatively efficient compared to the other Kyoto trading schemes, with a cost of 12 times the temperature-minimum path. The other Kyoto emissions limitations scenarios have costs of between 43 and 177 times the temperature-minimum path. Note that there is relatively little gain from trading within the OECD countries alone; most of the gain from Annex I trade arises from the inclusion of Russia and other Eastern European countries under the trading umbrella.

Figure 8.6 shows the impact of different strategies on the time path of world output. In the cases that allow large-scale trading, the overall impact of the Kyoto Protocol is relatively modest, between 0.1 and 0.2 percent of income per year.

Tables 8.6 and 8.7 and figure 8.7 show the economic impacts on different regions. Abatement costs in table 8.6 include net purchases of emissions permits. The net economic impact of a policy in table 8.7 is the environmental benefits of the policy minus the abatement costs (see chapter 7, section three, the first subsection for further description).

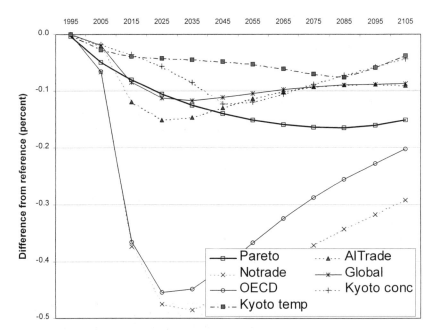

**Figure 8.6**
Impact of policy on world GDP

**Table 8.6**
Abatement costs in different regions for different policies (billions of 1990 U.S. dollars)

| | Base | Optimal | Kyoto concentration | Kyoto temperature | Global | AI trade | OECD | No trade |
|---|---|---|---|---|---|---|---|---|
| USA | 0 | 15 | 1 | 0 | 91 | 325 | 710 | 852 |
| OHI | 0 | 1 | -1 | -1 | 7 | 24 | 16 | 39 |
| Europe | 0 | 1 | -1 | -1 | 0 | -5 | -93 | 15 |
| EE | 0 | 12 | 2 | 1 | -28 | -112 | -2 | -2 |
| MI | 0 | 5 | 0 | 0 | -2 | -3 | -4 | -4 |
| LMI | 0 | 23 | 4 | 3 | -2 | -4 | -5 | -5 |
| China | 0 | 14 | 2 | 0 | -4 | -3 | -4 | -4 |
| LI | 0 | 27 | 5 | 3 | -2 | -5 | -7 | -8 |
| Annex I | 0 | 29 | 1 | -1 | 70 | 232 | 630 | 905 |
| ROW | 0 | 69 | 11 | 7 | -11 | -15 | -20 | -21 |
| World | 0 | 98 | 12 | 5 | 59 | 217 | 611 | 884 |

Note: Table 8.6 shows the discounted global costs of different targets or control strategies. The estimates are the discounted consumption in the base case minus the discounted consumption in the case in question, where these have been calculated excluding the environmental benefits of controls. Revenues from permit trading are implicit in the calculations.

**Table 8.7**
Net economic impacts in different regions for different policies (billions of 1990 U.S. dollars)

| | Base | Optimal | Kyoto concentration | Kyoto temperature | Global | AI trade | OECD | No trade |
|---|---|---|---|---|---|---|---|---|
| USA | 0 | 22 | 12 | 10 | -78 | -313 | -692 | -833 |
| OHI | 0 | 26 | 10 | 9 | 3 | -15 | -2 | -26 |
| Europe | 0 | 126 | 47 | 36 | 47 | 46 | 161 | 54 |
| EE | 0 | -9 | -1 | 0 | 29 | 113 | 3 | 3 |
| MI | 0 | 19 | 8 | 7 | 11 | 11 | 17 | 18 |
| LMI | 0 | 5 | 6 | 5 | 13 | 13 | 20 | 21 |
| China | 0 | -10 | 0 | 1 | 5 | 4 | 6 | 6 |
| LI | 0 | 20 | 12 | 9 | 19 | 21 | 32 | 34 |
| Annex I | 0 | 164 | 69 | 56 | 0 | -170 | -530 | -801 |
| ROW | 0 | 34 | 26 | 21 | 48 | 49 | 75 | 78 |
| World | 0 | 198 | 95 | 77 | 49 | -121 | -455 | -723 |

Note: Table 8.7 shows the total impacts of different targets or control strategies for different regions, including environmental benefits and net sales of emissions permits. The estimates are the difference between the present value of consumption in the case in question and the present value of consumption in the base. Positive values reflect net benefits while negative ones reflect net costs.

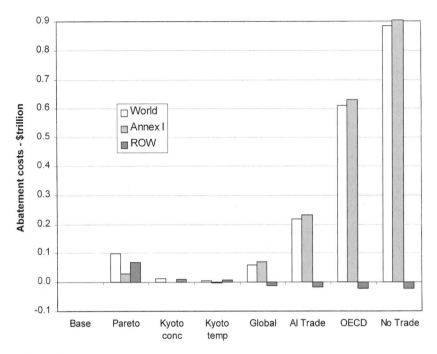

**Figure 8.7**
Regional impacts of alternative strategies

Table 8.6 indicates why the United States might be most unhappy with the Kyoto Protocol under which its cost is more than the global total for the Annex I version. The U.S. discounted cost is $325 billion, while the rest of the world has a net benefit (primarily through permit sales) of $108 billion. OECD Europe and OHI have very little net abatement costs in any versions of the protocol. The high cost of the Kyoto Protocol to the United States arises because $CO_2$ emissions are projected to grow much more rapidly in the United States than in other regions, so containing emissions is likely to prove much more expensive. (For a comparison of the baseline emissions, see figure 7.7.)

**Trading and transfers.** Permit sales and purchases are relatively small in the calculations in RICE-99, primarily because the impact of the Kyoto Protocol is relatively limited. In the global trade case, the only purchasers of permits are the high-income regions, which buy $6.3 billion worth of permits in 2015. In the Annex I trading case, the major trade takes place between the United States and R&EE; the transfer from the United States averages $22 billion per year over the next century. This is far less than many other estimates of the impact of the Kyoto Protocol.[7]

By examining the flows of permit revenues, we see two major flaws in the design of the Kyoto Protocol. First, the protocol caps the emissions of one group of countries at historical levels but does not do so for the non-Annex I countries. Bringing non-Annex I countries in under the additionality criterion assigns *reference* emissions to non-Annex I countries, thus giving them substantially different treatment from Annex I countries. A second and related major design flaw is assigning historical emissions. This gives a major windfall to those countries that had inefficient energy systems (particularly Russia, Eastern Europe, and Germany after its reunification with East Germany). A better procedure would be a rolling emissions base, which would remove the advantages of inefficiency and also remove the difference of treatment of non-Annex I and Annex I countries.

**Summary of economic impacts.** The overall impacts of the Kyoto Protocol and variants are complex, but the major points to emphasize are the following:

1. There are big impacts of virtually all variants of the Kyoto Protocol on the United States. As shown in table 8.6, the discounted value of production costs (exclusive of climate damages) range from $325 billion in the Annex I trading case to $852 billion in the no-trading case. Introducing global trading reduces abatement costs by about two-thirds, but it is probably unrealistic as a policy at present.

2. If damages are excluded, the major beneficiaries of the Kyoto Protocol are the regions with permits to sell. In the Annex I trading case, Russia and Eastern Europe gain $112 billion more in present value from permit sales than they incur in abatement costs. In the OECD trading case, OECD Europe is the major beneficiary.

3. The major beneficiary of the environmental effects of reducing emissions is Europe. The net economic impact on OECD Europe is positive in all experiments considered in this chapter, with the environmental benefits ranging from $35 to $127 billion.

4. Trading significantly reduces the aggregate cost of abatement, particularly trading with Russia and low-income countries like China.

---

7. See the different estimates collated in Weyant 1999.

But the counterpart of these efficiency gains is transfers—mainly from the United States—although they are not as large as indicated in other models.

## Costs and Damages

The focus has been primarily on the abatement costs, but it is always important to keep in mind that the point of reducing emissions is to reduce future damages. Our estimates indicate that there are likely to be substantial costs of global warming in any of the cases examined here; the discounted value of damages in the base case are approximately $4 trillion in present value.

The impact of different policies on both costs and damages is shown in figure 8.8 and table 8.8. The second set of bars shows the discounted value of the reduced damages. This figure shows that the policies reduce (discounted) damages by only a modest amount—between $100 and $300 billion out of total damages of $4 trillion. The maximum damage reduction from the Kyoto Protocol is $160 billion.

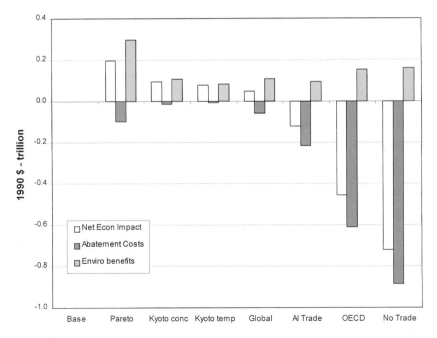

**Figure 8.8**
Regional impacts of alternative strategies

**Table 8.8**
Benefits, costs, and benefit-cost ratio of different approaches (costs and damages in billions of 1990 U.S. dollars)

|  | Base | Optimal | Kyoto concentration | Kyoto temperature | Global | AI trade | OECD | No trade |
|---|---|---|---|---|---|---|---|---|
| Abatement costs | 0 | 98 | 12 | 5 | 59 | 217 | 611 | 884 |
| Reduction in damages | 0 | 296 | 107 | 83 | 108 | 96 | 155 | 161 |
| Benefit-cost ratio | na | 3.02 | 9.07 | 15.22 | 1.82 | 0.44 | 0.25 | 0.18 |

Note: Calculations show the abatement costs and reduction of climate damages from alternative policies. Costs and damages are in billions of dollars discounted to 1995 in 1990 U.S. dollars.

Many readers may express surprise about the small impact of the Kyoto Protocol on the damages from global warming. The reasons are that, because there is so much inertia in the climate system and because the protocol does not limit the emissions of developing countries, the Kyoto Protocol reduces the global temperature increase by only a fraction of a degree over the next century. The other point shown in figure 8.8 and table 8.8 is that where- and when-inefficiency raise the costs of abatement substantially with little or no improvement in benefits. For example, moving from no controls to the Kyoto Protocol plan with Annex I trading incurs discounted abatement cost of $217 billion; however, the discounted value of damages decreases by only $96 billion. Moving from the Annex I version to the no-trade version increases benefits by $65 billion while increasing costs by $667 billion.

Finally, figure 8.9 shows the distribution of net impacts (including transfers and climate damages) of the seven major policies considered here.

The main conclusions that come from an examination of damages are that there are likely to be substantial damages from climate change,

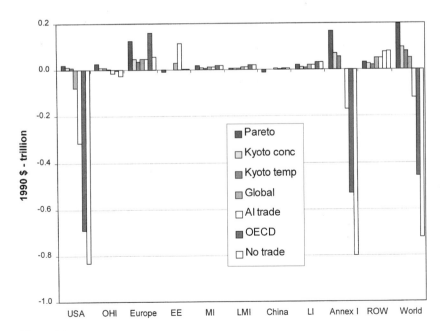

**Figure 8.9**
Net economic impact by region

but that the Kyoto Protocol does virtually nothing to mitigate the damages.

## Gains from Trade

Much has been made about the gains from trade. Table 8.5 indicates that the cost-effectiveness differs greatly across different options. We estimate that the minimum costs of attaining the environmental objectives of the Annex I version of the Kyoto Protocol are approximately $12 billion for the concentration objective and $5 billion for the temperature objective. By this standard, the Annex I protocol has costs that are 18 to 43 times the cost-effective amount, while the no-trade version has costs that are 74 to 177 times the cost-effective standard.

The cost savings from the minimum-cost variants over the versions of the Kyoto Protocol derive from when-efficiency and where-efficiency. Like the global trade run, the Kyoto temperature run allows emissions reductions to occur where they are least costly; but the Kyoto temperature run also has the additional flexibility of allowing the emissions reductions to take place when they are cost-effective, whereas all the Kyoto cases arbitrarily specify the time path of emissions reductions. When-efficiency accounts for about 6 percent of the difference in abatement costs between the minimum cost and the no trade variant; the remaining 94 percent, captured by the global trading case, is due to gains from trade, or where-efficiency. Of the gains from trade, 81 percent can be captured through Annex I trading.

One major surprise is that trading within the OECD only attains relatively little—33 percent—of the potential gains from trade. This low fraction comes about because the energy efficiencies are relatively similar within the large OECD blocs.

While targeting temperature directly produces the minimum-cost way of achieving the environmental objective in RICE, targeting concentrations is nearly as good; this policy captures 99 percent of the gains from when- and where-efficiency.

## Findings and Conclusions

This chapter examines the implications of the Kyoto Protocol and variants of that policy in a new integrated-assessment model of climate change and the world economy. Before moving to the major conclusions, it must be emphasized that these results should be taken with suitable reservations reflecting the difficulties inherent in the subject and the fact that this is but one of many models that can be used to estimate the impacts of the Kyoto Protocol. For instance, analysis of the Kyoto Protocol is inherently highly uncertain because of the nature of the protocol. The outcomes for countries depend upon the difference between emissions and a fixed historical benchmark. Small changes in growth rates can make big differences in the outcomes. A modest change in the projected growth rate of output, with carbon intensity constant, can make a substantial difference in the projected costs and benefits over the next few decades. This uncertainty is reflected in the differences across models or even across different versions of the same model.

This analysis concludes with three observations. First, the strategy behind the Kyoto Protocol has no grounding in economics or environmental policy. The approach of freezing emissions at a given level for a group of countries is not related to a particular goal for concentrations, temperature, or damages, nor does it bear any relation to an economically oriented strategy that would balance the costs and benefits of greenhouse-gas reductions. The emissions and concentrations implicit in the Kyoto Protocol are close to those in the optimal policy for the first few decades, but they are too low relative to the efficient policy over the next century. The basic difficulty is that the Kyoto Protocol targets only the emissions of high-income countries, which are likely to be a dwindling fraction of global emissions.

Second, while the environmental damages from climate change do not differ markedly among the different Kyoto Protocol variants, the costs of implementation vary enormously. The cost of the no-trade variant of the Kyoto Protocol is about fifteen times the cost of the global-trade variant. Most of the gains from trade come from including non-OECD countries such as Russia, China, and India. The costs of an efficiently designed Kyoto Protocol with global trading are $59 billion, while the benefits of the emissions reduction from the Kyoto Protocol are around $108 billion, for a benefit-cost ratio of 1.8. The benefit-cost ratio of the Annex I version is 0.44, however, while the ratio for the no-trade version is 0.18. Moreover, these estimates are based on the optimistic assumption that policies are efficiently implemented. The differences among implementation strategies emphasize the point that efficient design of the policy should be a major concern of policymakers.

Finally, the Kyoto Protocol has significant distributional consequences. Annex I countries pay the costs of protocol. These costs will
come either through abatement activities or through purchase of
permits. The lion's share of these costs are borne by the United States.
Indeed, the United States is a net loser while the rest of the world on
balance benefits from the Kyoto Protocol.

# 9       Managing the Global Commons

The final chapter summarizes the approach taken in this book and reviews the major conclusions. It begins with a summary of the new models discussed here along with a comparison between this model and other integrated assessment models. The chapter then summarizes the major findings from this study, and it concludes with some reflections on how societies can use the results of this and other models for the thorny problems involved in coping with the threat of global warming.

## Background

Global warming has become the major environmental policy issue of today. Concerns about the impact of global warming have increasingly been the subject of research and debate among natural and social scientists, and nations have undertaken a controversial new approach to curbing global environmental threats in adopting the Kyoto Protocol of December 1997. Presidential candidate Al Gore has called global warming one of the major global issues of the twentieth century.

Dealing with complex scientific and economic issues has increasingly involved developing scientific and economic models. Such models help analysts and decision makers unravel some of the complex interactions and understand future outcomes as well as the implications of alternative policies. In the economic literature, the DICE model developed in the early 1990s by Nordhaus (see especially Nordhaus 1994b) was one of the early IA models of the economics of climate change. This model developed an approach that links together from end to end the different facets of global warming. This book presents a newly elaborated model, RICE-99, along with its aggregated companion, DICE-99. These models are fully revised representations of the economics of global warming that build upon earlier work by the author and collaborators.

The purpose of this book is to integrate the economic aspects of emissions of greenhouse gases and damages from climate change with current scientific knowledge of the dynamics of climate change. The book provides a full description of the methodology as well as an analysis of alternative approaches to climate-change policy.

The analysis is laid out in eight chapters. The first chapter gives a general introduction to the subject. The following chapter presents the RICE-99 model, starting with a verbal description and following with a list of the equations. Chapter 3 describes the methods and data used to calibrate the model, and chapter 4 provides special detail on the calibration of the damages from climate change. Chapter 5 presents DICE-99 and describes its relation to RICE-99. Chapter 6 describes how the model is solved.

The next two chapters present the major results and some of the important conclusions. Chapter 7 describes the baseline scenario along with a number of alternative policies. Chapter 8 presents an economic analysis of the Kyoto Protocol. The appendixes provide a summary listing of the equations, a variable list, other summary tables, and the computer programs for the different models.

## Summary of the Model and Analysis

### The RICE/DICE-99 Models

An overview of the approach taken here in analyzing the economics of climate change is one that considers the climate change problem as an *economic* problem. It requires reducing potential consumption of goods and services today to reduce the damages and risks of climate change in the future. By taking costly steps to slow emissions of GHGs today, the economy reduces the amount of output that can be devoted to consumption and productive investment. The return for this climate investment is lower damages and therefore higher consumption in the future. The purpose of the book is to examine the major trade-offs involved in climate-change policy and to evaluate the relative efficiency of different policies.

In RICE-99, the world is composed of sovereign regions, represented by large countries (like the United States or China) or large regions (like OECD Europe or low-income countries). Each region is assumed to have a well-defined set of preferences over current and future consumption, described by a regional social welfare function that applies a pure time preference factor to the utility of future generations. Regions are assumed to maximize the social-welfare function subject to a number of economic and geophysical constraints. Society can select among alternative outcomes by choosing different savings rates and different emissions of GHGs.

The model contains both a traditional economic sector, similar to that found in many economic models, and a geophysical module designed for climate-change modeling. In the economic sectors, each region is assumed to produce a single commodity that can be used for either consumption or investment. Regions are not allowed to trade except to exchange goods for rights to carbon emissions.

Each region is endowed with an initial stock of capital and labor and an initial and region-specific level of technology. Population growth and technological change are exogenous, while capital accumulation is determined by each region's optimizing the flow of consumption over time. RICE-99 defines a new input into production called *carbon-energy*. Carbon-energy consists of the energy services derived from the consumption of carbon fuels, where carbon fuels are measured in carbon units. $CO_2$ emissions are therefore a joint product of using energy for productive purposes. The economy can reduce emissions through substitution of capital and/or labor for carbon-energy. Technological change takes two forms: economy-wide technological change and carbon-saving technological change. More precisely, economy-wide technological change is Hicks-neutral, while carbon-saving technological change is assumed to increase the ratio of energy services to $CO_2$ emissions.

The energy-related elements in the production function are calibrated using existing data on carbon emissions, energy use, energy prices, and energy-use price elasticities. These allow an empirically based carbon-reduction cost function, whereas most current integrated assessment models make reasonable but not empirically based specifications of the cost schedule. In RICE-99, a carbon supply curve is introduced which demonstrates that the costs of producing carbon-energy rise as cumulative extraction increases. Because the model employs the optimal-growth framework, fossil fuels are efficiently allocated, which implies that low-cost resources have scarcity rents (Hotelling rents) and that the scarcity rents on carbon-energy rise over time.

The environmental part of the model contains a number of geophysical relationships that link together the different factors affecting climate change. This part contains a carbon cycle, a radiative forcing equation, climate-change equations, and a climate-damage relationship. The geophysical sectors are simplified representations of more complex models. Although they have been built on first principles, our research shows that they closely track more elaborate models.

In the new models, endogenous emissions are limited to industrial $CO_2$ (which, as noted above, are a joint product of using carbon-energy). Other contributions to global warming are taken as exogenous. The new models contain a new structural approach to carbon-cycle modeling that uses a three-reservoir model calibrated to existing carbon-cycle models. Climate change is represented by global mean surface temperature, and the relationship uses a midrange climate sensitivity and a lag derived from coupled ocean-atmospheric models.

Understanding the economic impacts of climate change continues to be the thorniest issue in climate-change economics. This book presents a new synthesis of damage estimates. The analysis also presents new estimates for thirteen major regions rather than for the United States alone, although in the current version of RICE these are aggregated into eight regions. The new study focuses more heavily on the nonmarket aspects of climate change with particular importance given to the potential for abrupt or catastrophic risk. This approach is taken because the weight of the evidence indicates that the economic impacts of climate change on the market sectors of high-income countries are likely to be relatively limited. The major result is that impacts are likely to differ sharply by region. We estimate that Russia and other high-income countries (principally Canada) will benefit slightly from a modest global warming, while low-income regions—particularly Africa and India—and Europe appear to be quite vulnerable to climate change. The United States appears to be less vulnerable to climate change than many countries.

Some unfinished business should be noted. We reiterate that the damage function, particularly the response of developing countries and natural ecosystems to climate change, is poorly understood at present. An important open issue is the possibility of abrupt climate change; this is a central concern because, whereas scientists have improved their understanding of many elements of climate change, the potential for abrupt or catastrophic climatic change, for which precise mechanisms and probabilities have not been determined, cannot currently be ruled out. A related issue is that this book abstracts from issues of uncertainty, in which risk aversion and the possibility of learning may modify the stringency and timing of control strategies. Additionally, the calculations omit the interactions between climate change and other potential market failures, such as air pollution, taxes, and research and development, which might reinforce or weaken the logic behind greenhouse-gas reduction or carbon taxes. Although the model assumes substantial future technological change—both overall and carbon saving—it omits endogenous technological change. Finally, the model assumes that policies are efficiently implemented, which is undoubtedly an optimistic assumption given shortcomings in most environmental policies. These are all topics for further study.

### Differences between RICE-DICE and Other Models

The RICE-DICE-99 models are but one family in a growing population of IA models of the economics of global warming. What are the major differences among the models and what are their similarities? This is an enormous question, but a few general comments are worth making here.

First, recent surveys of IA models tend to classify them into two general categories: optimization and policy evaluation. The optimization models, of which DICE is an example, are ones that have a well-defined objective and can determine optimal policies. The policy evaluation models are ones that are more descriptive in nature and trace out different scenarios rather than attempt to identify the best policies.[1] While DICE continues in the optimization framework, the regional DICE-type models, such as RICE-99, are more of a hybrid of the two approaches. RICE-99 contains much descriptive information and, particularly in the spreadsheet versions, can easily be used as a policy-evaluation model. At the same time, because all welfare changes, including reductions in climate damages, are modeled as consumption changes, it can be used as an optimization model.

Second, the major difference between the DICE-RICE family and most other major IA models is that DICE-RICE contains a complete evaluation of the societal impacts or damages from climate change while most other models stop short of incorporating damages. Because damage estimates are so uncertain, their inclusion adds considerable uncertainty to that part of the model. On the other hand, to omit considerations of damages is to lose sight of the fundamental objective of

---

1. Excellent recent surveys of IA models are contained in chapter 10 in IPCC 1996c and Kolstad 1998.

climate change policy, which is to keep greenhouse-gas concentrations below dangerous levels, as stated in the FCCC. While the definition of the "dangerous levels" is still open to debate, it is important to keep the ultimate objectives in mind when considering alternative policies.

Third, each modeler has a special appreciation for his or her model's strengths and weaknesses. The major strengths of the DICE-RICE-99 models are two: first, they have been designed for transparency and ease of use and adoption by a wide range of people from students to researchers at the frontiers of their disciplines; and, second, the components are designed to reflect the state of the art in each area while maintaining a parsimonious representation. For example, RICE-99 contains a simple but reasonably accurate representation of the current state of knowledge about economies of different regions for doing long-run policy analysis. The modeling philosophy is that the DICE-RICE-99 models should be like a well-designed car—the parts and the whole are all optimized to get modelers where they want to go at low cost, given the prices of time, energy, and ability.

Finally, we would also like to describe some of the weaknesses of the DICE-RICE models relative to other IA models. Other models are far better for specialized tasks. For example, no sensible economist would *ever* use the production sector in these models to consider the role of business cycles or to make short-run forecasts. The energy sector in the DICE-RICE models is designed for global warming economics and is poorly served to analyze interfuel substitution. International trade is omitted. For these and similar issues that need finer detail, there are specialized models that can provide much better resolution.

**Major Results**

This book contains many results that have been reported in earlier chapters. Five important conclusions will be highlighted in this summary.

The first major point is that an efficient climate-change policy would be relatively inexpensive and would slow climate change surprisingly little. Our estimate is that the present value of global abatement costs for the optimal policy would be around $100 billion, or an annualized cost of about $5 billion per year. Another interesting result is that a short delay in implementing an optimal policy has little cost; indeed, it can cost far less than implementing an inefficient policy. (Recall that all dollar values in the text, tables, and graphs represent 1990 U.S.

dollars. Prices for the year 2000 are approximately 25 percent higher using the U.S. GDP deflator.)

The optimal policy reduces the global temperature rise to 2.34°C in 2100 and to 3.65°C in 2200. More stringent policies are ones that limit $CO_2$ concentrations to a doubling of preindustrial levels (which has present value cost of about $1,400 billion) and limiting global temperature increases to 2.5°C (costing $3,500 billion). Our estimate is that the optimal policy has discounted benefits of reduced damages of about $300 billion for a benefit-cost ratio of 3. The other two environmentally oriented policies have discounted benefits of about $700 billion (for concentrations limitation) and $1,100 billion (for limiting temperature increase to 2.5°C) for benefit-cost ratios one-half and one-third, respectively.

Second, some prominent policy proposals look highly inefficient. From bad to worst we would rank Kyoto (Annex I trading), Kyoto (OECD trading), limiting $CO_2$ concentrations to twice preindustrial levels, Kyoto (no trading), limiting climate change to a 2.5°C temperature rise, stabilizing global emissions at 1990 levels, and limiting climate change to a 1.5°C temperature rise (see table 7.2). As currently estimated, none of the policies except geoengineering has major net economic benefits. The most beneficial control option has a net benefit of only $200 billion in present value. On the other hand, inefficient policies can do significant economic damage.

The third point refers to findings about carbon taxes or permit prices. With respect to current climate-change policies, perhaps the most important finding is that the optimal carbon tax in the near term is in the $5 to $10 per ton range. As table 7.5 and figure 7.2 show, that price range is an appropriate target for policy for the next decade or so. Surprisingly, the environmentally oriented concentrations limits and 2.5°C temperature limits have relatively low carbon taxes: $4 and $12 per ton carbon in 2005, respectively. Policies that have near-term carbon taxes in the $100 per ton range, such as those associated with the Kyoto Protocol, are almost sure to fail a cost-benefit test because they impose excessive near-term abatement. Moreover, all policies that pass a cost-benefit test have near-term carbon taxes less than $15 per ton.

The fourth point concerns the revised view of the threat from global warming. The present study paints a less alarming picture of future climate change than other studies performed in the early 1990s. Whereas many studies projected baseline global temperature increases by 2100 in the 3 to 4°C range, a better reference estimate today would be close to 2.4°C warming in 2100. It is interesting to compare the results of the new model with the earlier DICE model. The optimal carbon tax and control rate in the early periods in the two models are very close; however, RICE-99 has significantly slower growth in emissions, concentrations, and other greenhouse-gas forcings. The slower buildup of concentrations, along with the evidence of the cooling effect of other gases and the phaseout of the CFCs, implies that the baseline global temperature increase for 2100 is 2.42°C in RICE-99 as compared to 3.28°C in the original DICE model. In addition, the new RICE-99 model has higher controls than the original DICE model. Hence the optimized global temperature increase in 2100 is 2.34°C in RICE-99 compared to 3.10°C in the original DICE model.

The final point is that an environmentally benign geoengineering policy would be highly beneficial. We estimate that a technological solution that would costlessly offset the climatic impacts of increasing greenhouse-gas concentrations would have a benefit of around $4 trillion in present value. This point is important because of the finding that the optimal policy in RICE-99 does not slow temperature change much over the next century. It is important to understand that this result comes about because of the costs of slowing climate change, not because climate damages are negligible.

We conclude that, although not an environmentally correct policy, geoengineering is a policy option that deserves more careful study and consideration. It has important advantages over the house-to-house combat of emissions reductions. One important advantage is that geoengineering appears to be inexpensive. Another feature is that it does not require near-unanimous agreement among all major countries to have an effective policy; indeed, the United States could easily undertake geoengineering by itself if other countries would give their assent. Given the controversies surrounding climate change, given the slow pace of effective international agreements, and particularly given the increasing concerns about potentially catastrophic impacts, it is clear that much more attention should be devoted to geoengineering options.

## Analysis of the Kyoto Protocol

The analysis of efficient paths is in one sense the analysis of policy in a vacuum—a vacuum in which powerful interest groups, disagreements among nations, incompetence and ignorance of policymakers, and inefficient implementation are all absent. A more realistic analysis would look at the Kyoto Protocol, which is the agreed-upon (but unratified) international agreement on how nations will begin to slow global warming. The analysis in chapter 8 has discovered a few key points.

First, we conclude that the Kyoto Protocol has no economic or environmental rationale. The approach of freezing emissions for a subgroup of countries is not related to a particular goal for concentrations, temperature, or damages. Nor does it bear any relation to an economically oriented strategy that would balance the costs and benefits of greenhouse-gas reductions.

Second, it is useful to compare the Kyoto Protocol with our estimates of the optimal policy. The carbon prices in the global version of the Kyoto Protocol are close to our estimates of optimal policy in the first few decades. The global emissions targets embodied in the Kyoto Protocol are close to those in the optimal policy for the first few decades. In the longer run, however, the emissions reductions targeted under the Kyoto Protocol are too low relative to the efficient policy. The basic difficulty is that the Kyoto Protocol targets only the emissions of high-income countries, which are likely to be a dwindling fraction of global emissions.

Third, the cost-effectiveness of the Kyoto Protocol will depend crucially on how it is implemented. One extreme would be the global-trading version, where all nations enter into an efficient trading arrangement and policies are efficiently implemented. Our estimate is that this policy would be reasonably efficient over the next few years. The costs of an efficiently designed Kyoto Protocol are $59 billion, while the benefits of the emissions reduction from the Kyoto Protocol are around $108 billion, for a benefit-cost ratio of 1.8. The global-trading scenario is highly unlikely, however, for it assumes participation of nations that are unwilling and in some cases unable to participate.

At the other pole would be the case where there is no trading of emissions allowances across major regions, either because of a breakdown in the agreement or because the trading regime is prohibitively expensive. The cost of the no-trade variant of the Kyoto Protocol is about fifteen times the cost of the global-trade variant assuming efficient implementation. Even if there is trading among high-income (OECD) regions, the costs are likely to be near the no-trade case. The benefit-cost ratio for the no-trade version is 0.2. These calculations emphasize that efficient design of the policy should be a major concern of policymakers.

Finally, the Kyoto Protocol has significant distributional consequences. The United States bears most of the costs of implementing the current version of the Kyoto Protocol. These costs will come either through abatement activities or through purchase of permits. The United States is a net loser from all variants of the protocol, while other high-income countries and the rest of the world either break even or benefit from the Kyoto Protocol.

## Concluding Thoughts

This book will be comforting to some and outrageous to others. If there is a single message, it is that climate change is a complex phenomenon, unlikely to be catastrophic in the near term, but potentially highly damaging in the long run. But it is a threat that is best approached with warm hearts and cool heads rather than bleeding hearts or hot heads. Global warming is a serious concern. The best estimate here is that the present value of damages is around $4 trillion, so it is well worth the effort to reduce the damages if that can be accomplished at low cost. This analysis suggests that current efforts to slow global warming through the Kyoto Protocol pay a high price but accomplish little.

The models developed here indicate that the pace of global warming will be slightly slower, and the near-term damages will be marginally smaller, than had been found in other studies or in earlier versions of the DICE-RICE models. The slower pace of future climate change is a hopeful but cautionary note to end on. Perhaps the reader can rest more soundly with the current evidence that climate change in the coming century is unlikely to enter the catastrophic range, particularly if effective steps are taken to slow climate change. The size of the revisions in the projections in the last decade and the fact that they come from so many different sources, however, are reminders of the enormous uncertainties that society faces in understanding and coping with the climate-change problem. So while we may sleep more soundly at night, we must be vigilant by day for changes that might lead our globe off in more dangerous directions.

Sets:  Time periods  $t$ (1995 = 0, 2005 = 1, etc.)

Regions  $J$

Trading blocs  $b$

(A.1)  $W_J = \sum_t U[c_J(t), L_J(t)]R(t).$

(A.2)  $R(t) = \prod_{v=0}^{t}[1 + \rho(v)]^{-10}.$

$\rho(t) = \rho(0)\exp(-g^P t).$

(A.3)  $U[c_J(t), L_J(t)] = L_J(t)\{\log[c_J(t)]\}.$

(A.4)  $g^{pop}{}_J(t) = g^{pop}{}_J(0)\exp(-\delta^{pop}{}_J t).$

$L_J(t) = L_J(0)\exp\left(\int_0^t g_J^{pop}(t)\right).$

(A.5a)  $Q_J(t) = \Omega_J(t)\{A_J(t)K_J(t)^\gamma L_J(t)^{1-\beta_J-\gamma} ES_J(t)^{\beta_J} - c^E{}_J(t)ES_J(t)\}.$

(A.5b)  $ES_J(t) = \varsigma_J(t)E_J(t)$

$g^Z{}_J(t) = g^Z{}_J(0)\exp(-\delta^Z{}_J t)$

$\varsigma_J(t) = \varsigma_J(0)\exp\left(\int_0^t g_J^Z(t)\right).$

(A.6)  $g^A{}_J(t) = g^A{}_J(0)\exp(-\delta^A{}_J t)$

$A_J(t) = A_J(0)\exp\left(\int_0^t g_J^A(t)\right).$

(A.7)  $Q_J(t) + \tau_J(t)[\Pi_J(t) - E_J(t)] = C_J(t) + I_J(t).$

(A.7′)    $\tau_J(t) = \tau_b(t) \ \forall \ J \in b$

$$\sum_{J \in b} II_J(t) \geq \sum_{J \in b} E_J(t)$$

$$\sum_{J \in b} II_J(t) = \sum_{J \in b} E_J(t) \ \text{if} \ \tau_b > 0$$

$\tau_b(t) \geq 0.$

(A.8)     $c_J(t) = C_J(t)/L_J(t).$

(A.9)     $K_J(t) = K_J(t-1)(1-\delta_K)^{10} + 10 \times I_J(t-1)$

$K_J(0) = K_J{}^*.$

(A.10)    $c^E{}_J(t) = q(t) + markup^E{}_J$

(A.11)    $CumC(t) = CumC(t-1) + 10 \times E(t)$

$$E(t) = \sum_J E_J(t).$$

(A.12)    $q(t) = \xi_1 + \xi_2[CumC(t)/CumC*]^{\xi_3}.$

(A.13a)   $M_{AT}(t) = 10 \times ET(t-1) + \phi_{11}M_{AT}(t-1) + \phi_{21}M_{UP}(t-1)$

$LU_J(t) = LU_J(0)(1-\delta_l)^t$

$$ET(t) = \sum_J (E_J(t) + LU_J(t))$$

$M_{AT}(0) = M_{AT}{}^*.$

(A.13b)   $M_{UP}(t) = \phi_{22}M_{UP}(t-1) + \phi_{12}M_{AT}(t-1) + \phi_{32}M_{LO}(t-1).$

$M_{UP}(0) = M_{UP}{}^*.$

(A.13c)   $M_{LO}(t) = \phi_{33}M_{LO}(t-1) + \phi_{23}M_{UP}(t-1)$

$M_{LO}(0) = M_{LO}{}^*.$

(A.14)    $F(t) = \eta\{\log[M_{AT}(t)/M_{AT}{}^{PI}]/\log(2)\} + O(t)$

$O(t) = -0.1965 + 0.13465t \qquad t < 11$

$\qquad = 1.15 \qquad\qquad\qquad t > 10.$

(A.15a)   $T(t) = T(t-1) + \sigma_1\{F(t) - \lambda T(t-1) - \sigma_2[T(t-1) - T_{LO}(t-1)]\}$

$T(0) = T*.$

(A.15b)   $T_{LO}(t) = T_{LO}(t-1) + \sigma_3[T(t-1) - T_{LO}(t-1)]$

$T_{LO} = T_{LO}*.$

(A.16)    $D_J(t) = \theta_{1,J}T(t) + \theta_{2,J}T(t)^2.$

(A.17)    $\Omega_J(t) = 1/[1 + D_J(t)].$

# Appendix B: Equations of DICE-99 Model

(B.1) $\quad W = \sum_t U[c(t), L(t)]R(t).$

(B.2) $\quad R(t) = \prod_{v=0}^{t}[1 + \rho(v)]^{-10}$

$\rho(t) = \rho(0)\exp(-g^{\rho}t).$

(B.3) $\quad U[c(t), L(t)] = L(t)\{\log[c(t)]\}.$

(B.4) $\quad g^{pop}(t) = g^{pop}(0)\exp(-\delta^{pop}t)$

$L(t) = L(0)\exp\left(\int_0^t g^{pop}(t)\right).$

(B.5) $\quad Q(t) = \Omega(t)\left(1 - b_1(t)\mu(t)^{b2}\right)A(t)K(t)^{\gamma}L(t)^{1-\gamma}.$

(B.6) $\quad g^A(t) = g^A(0)\exp(-\delta^A t)$

$A(t) = A(0)\exp\left(\int_0^t g^A(t)\right).$

(B.7) $\quad \Omega(t) = 1/[1 + D(t)].$

(B.8) $\quad D(t) = \theta_1 T(t) + \theta_2 T(t)^2.$

(B.9) $\quad g^b(t) = g^b(0)\exp(-\delta^b t)$

$b_1(t) = b_1(t-1)/(1 + g^b(t))$

$b_1(0) = b_1^*.$

(B.10) $\quad E(t) = (1 - \mu(t))\sigma(t)A(t)K(t)^{\gamma}L(t)^{1-\gamma}.$

(B.11) $\quad g^{\sigma}(t) = g^{\sigma}(0)\exp(-\delta^{\sigma}_1 t - \delta^{\sigma}_2 t^2)$

$\sigma(t) = \sigma(t-1)/(1 + g^{\sigma}(t))$

$\sigma(0) = \sigma^*.$

(B.12) $\quad Q(t) + \tau(t)[\Pi(t) - E(t)] = C(t) + I(t).$

(B.13)    $II(t) = E(t)$.

(B.14)    $c(t) = C(t)/L(t)$.

(B.15)    $K(t) = K(t-1)(1-\delta_K)^{10} + 10 \times I(t-1)$

        $K(0) = K^*$.

(B.16)    $LU(t) = LU(0)(1-\delta_1)^t$

        $ET(t) = E(t) + LU(t)$.

(B.17a)   $M_{AT}(t) = 10 \times ET(t-1) + \phi_{11}M_{AT}(t-1)$

               $-\phi_{12}M_{AT}(t-1) + \phi_{21}M_{UP}(t-1)$

        $M_{AT}(0) = M_{AT}{}^*$.

(B.17b)   $M_{UP}(t) = \phi_{22}M_{UP}(t-1) + \phi_{12}M_{AT}(t-1) + \phi_{32}M_{LO}(t-1)$

        $M_{UP}(0) = M_{UP}{}^*$.

(B.17c)   $M_{LO}(t) = \phi_{33}M_{LO}(t-1) + \phi_{23}M_{UP}(t-1)$

        $M_{LO}(0) = M_{LO}{}^*$.

(B.18)    $F(t) = \eta\{\log[M_{AT}(t)/M_{AT}{}^{PI}]/\log(2)\} + O(t)$

        $O(t) = -0.1965 + 0.13465t \qquad t < 11$

              $= 1.15 \qquad\qquad\qquad t < 10$.

(B.19)    $T(t) = T(t-1) + \sigma_1\{F(t) - \lambda T(t-1) - \sigma_2[T(t-1) - T_{LO}(t-1)]\}$

        $T(0) = T^*$.

(B.20)    $T_{LO}(t) = T_{LO}(t-1) + \sigma_3[T(t-1) - T_{LO}(t-1)]$

        $T_{LO} = T_{LO}{}^*$.

# Appendix C: Variable List

Variables are listed in the order they appear in appendixes A and B.

## Exogenous Variables and Parameters

| Variable | Description | Units |
| --- | --- | --- |
| $L(t)$ | Population | Millions |
| $R(t)$ | Social time preference discount factor | Pure number |
| $\rho(t)$ | Social time preference discount rate | Rate per year |
| $g^{\rho}$ | Growth rate of $\rho(t)$ | Rate per decade |
| $\rho(0)$ | Initial social time preference discount rate | Rate per year |
| $g^{\text{pop}}(t)$ | Growth rate of population | Rate per decade |
| $g^{\text{pop}}(0)$ | Initial population growth rate | Rate per decade |
| $\delta^{\text{pop}}$ | Rate of decline of $g^{\text{pop}}(t)$ | Rate per decade |
| $L(0)$ | Initial population | Millions |
| $A(t)$ | Total factor productivity | Determined by units of inputs in production function |
| $\gamma$ | Elasticity of output with respect to capital | Pure number |
| $\beta$ | Elasticity of output with respect to carbon-energy | Pure number |

| | | |
|---|---|---|
| $\varsigma(t)$ | Ratio of carbon-energy to industrial carbon emissions | Pure number |
| $g^Z(t)$ | Growth rate of $\varsigma(t)$ | Rate per decade |
| $g^Z(0)$ | Initial growth rate of $\varsigma(t)$ | Rate per decade |
| $\delta^Z$ | Rate of decline of $g^Z(t)$ | Rate per decade |
| $\varsigma(0)$ | Initial ratio of carbon-energy to industrial carbon emissions | Pure number |
| $g^A(t)$ | Growth rate of $A(t)$ | Rate per decade |
| $g^A(0)$ | Initial productivity growth rate | Rate per decade |
| $\delta^A$ | Rate of decline of $g^A(t)$ | Rate per decade |
| $A(0)$ | Initial total factor productivity | Determined by units of inputs in production function |
| $\Pi(t)$ | Industrial carbon emissions permits | GtC per year |
| $\delta_K$ | Rate of depreciation of capital | Rate per year |
| $K^*$ | Initial capital stock | Trillions of 1990 dollars |
| markup$^E$ | Regional energy services markup | \$1,000 / ton carbon |
| $\xi_1, \xi_2, \xi_3$ | Parameters of long-run industrial emissions supply curve | Varies by parameter |
| CumC$^*$ | Point of diminishing returns in carbon extraction | GtC |
| $\phi_{11}, \phi_{12},$ $\phi_{21}, \phi_{22},$ $\phi_{23}, \phi_{32},$ $\phi_{33}$ | Parameters of carbon transition matrix | Pure number |
| LU$(t)$ | Land-use carbon emissions | GtC per year |
| LU$(0)$ | Initial land-use carbon emissions | GtC per year |
| $\delta_l$ | Rate of decline of land-use emissions | Rate per decade |
| $M_{AT}^*$ | Initial atmospheric concentration of $CO_2$ | GtC |
| $M_{UP}^*$ | Initial concentration of $CO_2$ in upper oceans/biosphere | GtC |

| | | |
|---|---|---|
| $M_{LO}^*$ | Initial concentration of $CO_2$ in deep oceans | GtC |
| $\eta$ | Increase in radiative forcing due to doubling of $CO_2$ concentrations from preindustrrial levels. | $W/m^2$ |
| $M_{AT}^{PI}$ | Preindustrial $CO_2$ concentration | GtC |
| $O(t)$ | Increase in radiative forcing over preindustrial levels due to exogenous anthropogenic causes | $W/m^2$ |
| $\sigma_1, \sigma_2, \sigma_3,$ | Temperature dynamics parameters | Varies by parameter |
| $\lambda$ | $\eta/\lambda$ is the climate sensitivity, or equilibrium increase in temperature due to $CO_2$ concentration doubling | |
| $T^*$ | Initial atmospheric temperature | °C over 1900 level |
| $T_{LO}^*$ | Initial ocean temperature | °C over 1900 level |
| $\theta_1, \theta_2$ | Parameters of damage function | Varies by parameter |

### Specific to DICE-99

| | | |
|---|---|---|
| $b_1(t)$ | Coefficient on control rate in abatement cost function in DICE-99 | Pure number |
| $b2$ | Exponent on control rate in abatement cost function in DICE-99 | Pure number |
| $g^b(t)$ | Rate of decline of $b_1(t)$ | Rate per decade |
| $g^b(0)$ | Initial rate of decline of $b_1(t)$ | Rate per decade |
| $\delta^b$ | Rate of decline of $g^b(t)$ | Rate per decade |
| $b_1^*$ | Initial value of $b_1(t)$ | Pure number |
| $\sigma(t)$ | Base-case ratio of industrial emissions to output in DICE-99 | Tons / $1,000 |
| $g^\sigma(t)$ | Rate of decline of $\sigma(t)$ | Rate per decade |
| $g^\sigma(0)$ | Initial rate of decline of $\sigma(t)$ | Rate per decade |

| $\delta^\sigma_1, \delta^\sigma_2$ | Parameters that determine rate of decline of $g^\sigma(t)$ | Pure numbers |
|---|---|---|
| $\sigma^*$ | Initial value of $\sigma(t)$ | Tons / $1,000 |

## Endogenous Variables—RICE-99

| W | Welfare | Utils |
|---|---|---|
| U(t) | Utility during period t | Utils |
| c(t) | Per capita consumption | $ Million per year per person |
| Q(t) | Output | per year Trillions of 1990 dollars per year |
| $\Omega(t)$ | Climate-change damage factor on gross output | Pure number |
| K(t) | Capital stock | Trillions of 1990 dollars |
| ES(t) | Carbon-energy; energy services from carbon fuels | GtC per year |
| $c^E(t)$ | Cost of carbon-energy | $1,000 per ton |
| E(t) | Industrial $CO_2$ emissions | GtC per year |
| $\tau(t)$ | Industrial emission permit price | $1,000 per ton |

Note: Could also treat permit price as exogenous, in which case allocation of permits are constrained by (2.7′) (see chapter 2, part four)

| C(t) | Consumption | Trillions of 1990 dollars per year |
|---|---|---|
| I(t) | Investment | Trillions of 1990 dollars per year |
| $\tau_b(t)$ | Industrial emissions permit price in trading bloc b | $1,000 per ton |
| q(t) | Cost of extraction of industrial emissions | $1,000 per ton carbon |
| CumC(t) | Cumulative industrial carbon emissions | GtC |
| E(t) | World industrial carbon emissions | GtC per year |

| $M_{AT}(t)$ | Atmospheric $CO_2$ concentration | GtC |
|---|---|---|
| $ET(t)$ | World total $CO_2$ emissions | GtC per year |
| $M_{UP}(t)$ | Upper oceans/biosphere $CO_2$ concentration | GtC |
| $M_{LO}(t)$ | Lower oceans $CO_2$ concentration | GtC |
| $F(t)$ | Radiative forcing, increase over preindustrial level | $W/m^2$ |
| $T(t)$ | Atmospheric temperature, increase over 1900 level | °C |
| $T_{LO}(t)$ | Lower ocean temperature, increase over 1900 level | °C |
| $D(t)$ | Climate damage as fraction of net output | Pure number |

### Specific to DICE-99

| $\mu(t)$ | Industrial emission control rate | Pure number |
|---|---|---|

# Appendix D: GAMS Code for RICE-99, Base Case and Optimal Case

```
$OFFSYMXREF OFFSYMLIST
OPTION SOLPRINT=OFF;
OPTION DECIMALS=3;
OPTION NLP=MINOS5;
OPTION ITERLIM = 99999;
OPTION LIMROW = 0;
OPTION LIMCOL = 0;
OPTION RESLIM = 99999;

SETS   T   /1*35/
TFIRST(T)
TLAST(T)
TEARLY(T)
TLATE(T)
N            /USA, OHI, EUROPE, EE, MI, LMI, CHINA, LI/
ANNEXI(N)    /USA, OHI, EUROPE, EE/
ROW(N)       /MI, LMI, CHINA, LI/
POSPOP(N)    /USA, EE, MI, LMI, CHINA, LI/
ITER         /1*20/
REPS         /1*5/;

TFIRST(T)=YES$(ORD(T)=1);
TLAST(T)=YES$(ORD(T)=CARD(T));
TEARLY(T)=YES$(ORD(T) LE 20);
TLATE(T)=YES$(ORD(T) GE 21);
```

SCALARS

| | | |
|---|---|---|
| GAMMA | Capital share | /0.3/ |
| DELTA | Annual rate of depreciation (percent) | /10/ |
| CARBMAX | Point of diminishing returns in carbon extraction (GTC) | /6000/ |
| EXPCARB | Exponent on cost of extraction function | /4/ |
| LUGR | Rate of decline (percent per decade) in land use emissions | /10/ |
| M0 | Initial atmospheric concentration of $CO_2$ (GTC) | /735/ |
| MU0 | Initial concentration of $CO_2$ in upper box (GTC) | /781/ |
| ML0 | Initial concentration of $CO_2$ in deep oceans (GTC) | /19230/ |
| TE0 | Initial atmospheric temperature (deg C above preind) | /0.43/ |
| TL0 | Initial temperature of deep oceans (deg C above preind) | /0.06/ |
| SAT | Speed of adjustment parameter for atm. temperature | /0.226/ |
| CS | Equilibrium atm temp increase for $CO_2$ doubling (deg C) | /2.9078/ |
| HLAL | Coefficient of heat loss from atm to deep oceans | /0.44/ |
| HGLA | Coefficient of heat gain by deep oceans | /0.02/ |
| SRTP0 | Initial social rate of time preference (pct per year) | /3/ |
| SRTPGR | Rate of decline of srtp (pct per year) | /0.25719/ |
| Q | Utility derivative scaling factor | /10/ |

* Carbon cycle transition coefficients (percent per decade)

| | | |
|---|---|---|
| TRAA | Atmosphere to atmosphere | /66.616/ |
| TRUA | Upper box to atmosphere | /27.607/ |
| TRAU | Atmosphere to upper box | /33.384/ |
| TRUU | Upper box to upper box | /60.897/ |
| TRLU | Deep oceans to upper box | /0.422/ |

| TRUL | Upper box to deep oceans | /11.496/ |
| TRLL | Deep oceans to deep oceans | /99.578/ |
| WPDVC | World present value of consumption ($ billions) | |
| PDVCBASE | World present value of consumption in base ($ billions) | |
| TEI | Total economic impact of policy ($ billions); | |

SETS  OUTWELF Categories of data on welfare and production function

/DAM1, DAM2, K0, ALPHA, PHIGR, PHIGRGR/

| *Dam1 | Damage coefficient on temperature |
| *Dam2 | Damage coefficient on temperature squared |
| *K0 | Initial capital stock ($trill) |
| *Alpha | Elasticity of output with respect to carbon services |
| *Phigr | Initial growth rate of ratio of carbon emissions to carbon services (pct per dec) |
| *Phigrgr | Rate of decline of Phigr (pct per decade) |

EMPOPPRD Categories of data on emissions and population and productivity

/MU, LU0, L0, LGR, LGRGR, TFP0, TFPGR, TFPGRGR/

| *Mu | Markup on carbon ($ per tonne) |
| *Lu0 | Initial carbon emissions from land use change (GTC per year) |
| *L0 | Initial population (millions) |
| *Lgr | Initial population growth rate (pct per decade) |
| *Lgrgr | Rate of decline in population growth rate (pct per decade) |
| *Tfp | Initial total factor productivity |
| *Tfpgr | Initial productivity growth rate (pct per decade) |
| *Tfpgrgr | Rate of decline in productivity growth rate (pct per decade); |

TABLE OW(OUTWELF,N)

| | USA | OHI | EUROPE | EE |
|---|---|---|---|---|
| DAM1 | −0.0026 | −0.007 | −0.001 | −0.0076 |
| DAM2 | 0.0017 | 0.003 | 0.0049 | 0.0025 |
| K0 | 12.83 | 8.954 | 14.968 | 1.219 |
| ALPHA | 0.091 | 0.059 | 0.057 | 0.08 |

| PHIGR | −11.9 | −12.5 | −11.0 | −32.0 |
|---|---|---|---|---|
| PHIGRGR | 9.2 | 7.0 | 6.8777 | 15.0 |
| + | MI | LMI | CHINA | LI |
| DAM1 | 0.0039 | 0.0022 | −0.0041 | 0.01 |
| DAM2 | 0.0013 | 0.0026 | 0.002 | 0.0027 |
| K0 | 3.048 | 1.907 | 0.908 | 1.564 |
| ALPHA | 0.087 | 0.053 | 0.096 | 0.074 |
| PHIGR | −13.0 | −18.5 | −32.0 | −20.0 |
| PHIGRGR | 10.0 | 12.0 | 14.0 | 15.0; |

TABLE EPP(EMPOPPRD,N)

| | USA | OHI | EUROPE | EE |
|---|---|---|---|---|
| MU | 300.0 | 350.0 | 400.0 | −38.12 |
| LU0 | 0.0 | 0.0 | 0.0 | 0.0 |
| L0 | 260.71 | 190.75 | 383.4 | 342.14 |
| LGR | 9.0 | −1.0 | −6.0 | 1.3 |
| LGRGR | 32.417 | 26.5239 | 63.6195 | 13.6397 |
| TFP0 | 0.0996 | 0.0843 | 0.0748 | 0.02 |
| TFPGR | 3.8 | 3.9 | 4.1 | 11.0 |
| TFPGRGR | 1.5 | 0.5 | 1.5 | 4.0 |
| + | MI | LMI | CHINA | LI |
| MU | 250.0 | −2.63 | −41.09 | 18.78 |
| LU0 | 0.39645 | 0.21813 | 0.04094 | 0.472 |
| L0 | 313.08 | 564.82 | 1198.5 | 2379.15 |
| LGR | 16.0 | 20.0 | 8.52 | 24.0 |
| LGRGR | 24.7256 | 25.6632 | 25.3216 | 23.6399 |
| TFP0 | 0.0398 | 0.0172 | 0.0093 | 0.0078 |
| TFPGR | 8.0 | 11.0 | 15.0 | 12.0 |
| TFPGRGR | 3.0 | 4.5 | 5.0 | 4.0; |

PARAMETERS

| | |
|---|---|
| PHI(T,N) | Ratio of carbon emissions to carbon services |
| PHICGR(T,N) | Cumulative exponential growth rate of phi |
| LU(T,N) | Carbon emissions from land-use change (GTC per year) |

L(T,N)                    Population (millions)

LCGR(T,N)                 Cumulative exponential population growth rate

TFP(T,N)                  Total factor productivity

TFPCGR(T,N)               Cumulative exponential productivity growth rate

DAMCOEFF(T,N)   Damage coefficient in base case

PCYRATIO(T,N)    Ratio of per capita output to 1995;

```
PHICGR(T,N)=(OW("PHIGR",N)/OW("PHIGRGR",N))*(1–
    EXP(–(ORD(T)–1)*OW("PHIGRGR",N)/100));
LCGR(T,N)=(EPP("LGR",N)/EPP("LGRGR",N))*(1–
    EXP(–(ORD(T)–1)*EPP("LGRGR",N)/100));
LCGR(T, "OHI")=(EPP("LGR","OHI")/EPP("LGRGR","OHI"))*(1–
    EXP(–(ORD(T)–4)*EPP("LGRGR","OHI")/100));
LCGR(T, "EUROPE")=(EPP("LGR","EUROPE")/EPP("LGRGR",
    "EUROPE"))*(1–
    EXP(–(ORD(T)–4)*EPP("LGRGR","EUROPE")/100));
TFPCGR(T,N)=(EPP("TFPGR",N)/EPP("TFPGRGR",N))*(1–
    EXP(–(ORD(T)–1)*EPP("TFPGRGR",N)/100));
PHI(T,N)=EXP(PHICGR(T,N));
LU(T,N)=EPP("LU0",N)*(1–LUGR/100)**(ORD(T)–1);
L(T,POSPOP)=EPP("L0",POSPOP)*EXP(LCGR(T,POSPOP));
L("4", "OHI")=197.26;
L(T,"OHI")$(ORD(T)>4)=L("4", "OHI")*EXP(LCGR(T,"OHI"));
L("1", "OHI")=EPP("L0", "OHI");
L("2", "OHI")=199.03;
L("3", "OHI")=199.61;
L("4", "EUROPE")=381.1;
L(T,"EUROPE")$(ORD(T)>4)=L("4",
    "EUROPE")*EXP(LCGR(T,"EUROPE"));
L("1", "EUROPE")=EPP("L0", "EUROPE");
L("2", "EUROPE")=388.5;
L("3", "EUROPE")=386.1;
TFP(T,N)=EPP("TFP0",N)*EXP(TFPCGR(T,N));
```

PARAMETERS

| YEAR(T) | Year |
|---|---|
| EXOGFORC(T) | Exogenous forcing (W per meter squared) |
| SRTP(T) | Social rate of time preference (pct per year) |
| STPF(T) | Social time preference factor |
| RHOSTAR(T) | Unnormalized social welfare discount factor |
| SWDF(T) | Social welfare discount factor |
| WINDEM(T) | World industrial emissions (GtC per year) |
| SWDFBASE(T) | Social welfare discount factor in base case |
| CTAX(T) | Carbon tax ($ per metric ton) |

*** NAMES FOR DISCOUNT FACTOR USED IN ITERATIVE SEARCH ALGORITHM

RHOPREV(T)

SWDFPREV(T)

RHO(T);

YEAR(T)=1995+(ORD(T)–1)*10;

EXOGFORC(T)=(–0.1965+(ORD(T)–1)*0.13465)$(ORD(T)<12)+1.15$(ORD(T)>11);

SRTP(T)=100*(SRTP0/100)*EXP(–(SRTPGR/100)*10*(ORD(T)–1));

STPF("1")=1;

LOOP(T,
   STPF(T+1)=STPF(T)/((1+SRTP(T)/100)**10););

VARIABLES

| Y(T,N) | Output | ($trillions per year) |
|---|---|---|
| K(T,N) | Capital stock | ($trillions per year) |
| CA(T,N) | Industrial carbon emissions | (GTC per year) |
| C(T,N) | Consumption | ($trillions per year); |

VARIABLES

| E(T) | World carbon emissions (GTC per year) |
|---|---|
| M(T) | Atmospheric concentration of carbon (GTC) |
| MU(T) | Concentration of carbon in upper box (GTC) |
| ML(T) | Concentration of carbon in deep oceans (GTC) |

TE(T)      Atmospheric temperature (deg C above preindustrial)

F(T)        Radiative forcing (W per meter squared)

TL(T)       Deep ocean temperature (deg C above preindustrial)

CCA(T)    Cumulative industrial carbon emissions (GTC)

UTILITY   Social welfare function;

POSITIVE VARIABLES Y, K, CA, C, E, M, MU, ML, TE, F, TL;

CCA.LO(T)=0;

E.LO(T)=5;

M.LO(T)=M0;

ML.LO(T)=ML0;

MU.LO(T)=MU0;

F.LO(T)=4.1*LOG(M.LO(T)/596.4)/LOG(2)+EXOGFORC(T);

TE.LO(T)=TE0;

TL.LO(T)=TL0;

K.LO(T, N)=0.1;

Y.LO(T, N)=0.1;

CA.LO(T, N)=0.01;

CA.UP(T, N)=1000;

C.LO(T, N)=0.05;

***STARTING VALUES FOR BASE CASE

$INCLUDE STARTVAL

RHOSTAR(T)=SUM(N,   Y.L(T,N)*STPF(T)*L(T,N)/C.L(T,N))/SUM(N, Y.L(T,N));

SWDF(T)=RHOSTAR(T)/RHOSTAR('1');

DAMCOEFF(T,N)=1/(1+OW("DAM1",N)*SVT(T)+OW("DAM2",N)*SVT(T)**2);

***

EQUATIONS

YY(T,N)          Output

YYTEXOG(T,N)   Output with temperature constant

BC(T,N)          Budget constraint

KFIRST(T,N)     First period capital stock

KK(T,N)              Capital accumulation equation
CCTFIRST(T)          First period cumulative carbon
CCACCA(T)            Cumulative carbon emissions
EE(T)                World emissions
MFIRST(T)            First period atmospheric concentration
MM(T)                Atmospheric concentration
MUFIRST(T)           First period upper box concentration
MUMU(T)              Upper box concentration
MLFIRST(T)           First period deep ocean concentration
MLML(T)              Deep ocean concentration
TEFIRST(T)           First period temperature
TETE(T)              Atmospheric temperature
FF(T)                Radiative forcing
TLFIRST(T)           First period deep ocean temperature
TLTL(T)              Deep ocean temperature
OBJ                  Objective function;

YY(T,N)..   Y(T,N)=E=
(TFP(T,N)*(K(T,N)**GAMMA)*(L(T,N)**(1–GAMMA–
        OW("ALPHA",N)))
            *((CA(T,N)/PHI(T,N))**OW("ALPHA",N))–
((EPP("MU",N)+113+700*(CCA(T)/CARBMAX)**EXPCARB)/1000)*C
        A(T,N)/PHI(T,N))
            /(1+OW("DAM1",N)*TE(T)+OW("DAM2",N)*TE(T)**2);
YYTEXOG(T,N)..       Y(T,N)=E=
            (TFP(T,N)*(K(T,N)**GAMMA)*(L(T,N)**(1–GAMMA–
OW("ALPHA",N)))
            *((CA(T,N)/PHI(T,N))**OW("ALPHA",N))–
((EPP("MU",N)+113+700*(CCA(T)/CARBMAX)**EXPCARB)/1000)*C
        A(T,N)/PHI(T,N))*
            DAMCOEFF(T,N);
KFIRST(TFIRST,N)..   K(TFIRST,N)=E=OW("K0",N);
KK(T+1,N)..   K(T+1,N)=E=((1–DELTA/100)**10)*K(T,N)+10*(Y(T,N)–
            C(T,N));

```
BC(T,N)..   C(T,N)=L=Y(T,N);
CCTFIRST(TFIRST)..  CCA(TFIRST)=E=0;
CCACCA(T+1)..          CCA(T+1)=E=CCA(T)+10*SUM(N, CA(T,N));
EE(T)..    E(T)=E=SUM(N, CA(T,N)+LU(T,N));
MFIRST(TFIRST).. M(TFIRST)=E=M0;
MM(T+1)..
M(T+1)=E=10*E(T)+(TRAA/100)*M(T)+(TRUA/100)*MU(T);
MUFIRST(TFIRST)..   MU(TFIRST)=E=MU0;
MUMU(T+1)..
MU(T+1)=E=(TRAU/100)*M(T)+(TRUU/100)*MU(T)+(TRLU/100)*M
      L(T);
MLFIRST(TFIRST)..   ML(TFIRST)=E=ML0;
MLML(T+1).. ML(T+1)=E=(TRUL/100)*MU(T)+(TRLL/100)*ML(T);
TEFIRST(TFIRST)..   TE(TFIRST)=E=TE0;
TETE(T+1)..   TE(T+1)=E=TE(T)+SAT*(F(T)-(4.1/CS)*TE(T)-
                  HLAL*(TE(T)-TL(T)));
FF(T)..     F(T)=E=4.1*LOG(M(T)/596.4)/LOG(2)+EXOGFORC(T);
TLFIRST(TFIRST).. TL(TFIRST)=E=TL0;
TLTL(T+1)..    TL(T+1)=E=TL(T)+HGLA*(TE(T)-TL(T));
OBJ..        UTILITY=E=Q*SUM(T, SWDF(T)*SUM(N, C(T,N)));
***SOLVING BASE CASE
MODEL BASE/YYTEXOG, KFIRST, KK, BC, CCTFIRST, CCACCA,
            EE, MFIRST, MM,
               MUFIRST, MUMU, MLFIRST, MLML, TEFIRST, TETE, FF,
               TLFIRST, TLTL, OBJ/;
LOOP (ITER,
SWDF(T)=RHOSTAR(T)/RHOSTAR('1');
BASE.OPTFILE=1;
SOLVE BASE MAXIMIZING UTILITY USING NLP;
RHOPREV(T)=SUM(N, Y.L(T,N)*STPF(T)*L(T,N)/C.L(T,N))/SUM(N,
          Y.L(T,N));
SWDFPREV(T)=RHOPREV(T)/RHOPREV('1');
DISPLAY SWDF, SWDFPREV;
```

```
RHO(T)=RHOSTAR(T)+0.1*(RHOPREV(T)–RHOSTAR(T));
RHOSTAR(T)=RHO(T);
DAMCOEFF(T,N)=1/(1+OW("DAM1",N)*TE.L(T)+OW("DAM2",N)*
            TE.L(T)**2);
DISPLAY TE.L;);
SWDFBASE(T)=SWDF(T);
***
***BASE CASE OUTPUT
PCYRATIO(T,N)=(Y.L(T,N)/L(T,N))/(Y.L("1",N)/L("1",N));
WINDEM(T)=SUM(N, CA.L(T,N));
PDVCBASE=10*1000*UTILITY.L/Q;
DISPLAY PCYRATIO, WINDEM, PDVCBASE;
***
*** SOLVING OPTIMAL CASE
MODEL OPT /YY, KFIRST, KK, BC, CCTFIRST, CCACCA, EE,
            MFIRST, MM,
                MUFIRST, MUMU, MLFIRST, MLML, TEFIRST, TETE, FF,
                TLFIRST, TLTL, OBJ/;
LOOP (REPS,
SWDF(T)=RHOSTAR(T)/RHOSTAR('1');
OPT.OPTFILE=1;
SOLVE OPT MAXIMIZING UTILITY USING NLP;
RHOPREV(T)=SUM(N, Y.L(T,N)*STPF(T)*L(T,N)/C.L(T,N))/SUM(N,
            Y.L(T,N));
SWDFPREV(T)=RHOPREV(T)/RHOPREV('1');
DISPLAY SWDF, SWDFPREV;
RHO(T)=RHOSTAR(T)+0.1*(RHOPREV(T)–RHOSTAR(T));
RHOSTAR(T)=RHO(T););
***
*** OPTIMAL OUTPUT
PCYRATIO(T,N)=(Y.L(T,N)/L(T,N))/(Y.L("1",N)/L("1",N));
WINDEM(T)=SUM(N, CA.L(T,N));
CTAX(T)=–1000*EE.M(T)/YY.M(T, 'USA');
```

WPDVC=10*1000*SUM(T, SWDFBASE(T)*SUM(N, C.L(T,N)));
TEI=WPDVC–PDVCBASE;
DISPLAY PCYRATIO, WINDEM, CTAX, TE.L, WPDVC, TEI;
***

*** STARTVAL –STARTING VALUE PROGRAM

PARAMETER

CUMCARB(T)    Approximate cumulative carbon emissions (GTC);
CUMCARB(T)=90*(ORD(T)–1);

EQUATIONS

| | |
|---|---|
| YYU(T,N) | Output |
| BCU(T,N) | Budget constraint |
| KFU(T,N) | First period capital stock |
| KKU(T,N) | Capital accumulation equation |
| OBJU | Objective function |
| YYO(T,N) | Output |
| BCO(T,N) | Budget constraint |
| KFO(T,N) | First period capital stock |
| KKO(T,N) | Capital accumulation equation |
| OBJO | Objective function |
| YYE(T,N) | Output |
| BCE(T,N) | Budget constraint |
| KFE(T,N) | First period capital stock |
| KKE(T,N) | Capital accumulation equation |
| OBJE | Objective function |
| YYEE(T,N) | Output |
| BCEE(T,N) | Budget constraint |
| KFEE(T,N) | First period capital stock |
| KKEE(T,N) | Capital accumulation equation |
| OBJEE | Objective function |
| YYMI(T,N) | Output |
| BCMI(T,N) | Budget constraint |
| KFMI(T,N) | First period capital stock |

KKMI(T,N)   Capital accumulation equation

OBJMI       Objective function

YYLM(T,N)   Output

BCLM(T,N)   Budget constraint

KFLM(T,N)   First period capital stock

KKLM(T,N)   Capital accumulation equation

OBJLM       Objective function

YYC(T,N)    Output

BCC(T,N)    Budget constraint

KFC(T,N)    First period capital stock

KKC(T,N)    Capital accumulation equation

OBJC        Objective function

YYLI(T,N)   Output

BCLI(T,N)   Budget constraint

KFLI(T,N)   First period capital stock

KKLI(T,N)   Capital accumulation equation

OBJLI       Objective function;

YYU(T,"USA")..   Y(T,"USA") =E=

(TFP(T,"USA")*(K(T,"USA")**GAMMA)*(L(T,"USA")**(1–GAMMA–
   OW("ALPHA","USA")))*((CA(T,"USA")/PHI(T,"USA"))**OW("AL
   PHA","USA"))–((EPP("MU","USA")+113+700*(CUMCARB(T)/
   CARBMAX)**EXPCARB)/1000)*CA(T,"USA")/PHI(T,"USA"));

KFU(TFIRST,"USA")..   K(TFIRST,"USA")=E=OW("K0","USA");

KKU(T+1,"USA")..

K(T+1,"USA")=E=((1–DELTA/100)**10)*K(T,"USA")+
   10*(Y(T,"USA")–C(T,"USA"));

BCU(T,"USA")..   C(T,"USA")=L=Y(T,"USA");

OBJU..   UTILITY=E=SUM(T, STPF(T)*L(T,"USA")*
            LOG(C(T,"USA")/L(T,"USA")));

*OHI

YYO(T,"OHI")..   Y(T,"OHI")=E=

(TFP(T,"OHI")*(K(T,"OHI")**GAMMA)*(L(T,"OHI")**(1–GAMMA–
  OW("ALPHA","OHI")))*((CA(T,"OHI")/PHI(T,"OHI"))**OW("AL
  PHA","OHI"))–((EPP("MU","OHI")+113+700*(CUMCARB(T)/
  CARBMAX)**EXPCARB)/1000)*CA(T,"OHI")/PHI(T,"OHI"));

KFO(TFIRST,"OHI").. K(TFIRST,"OHI")=E=OW("K0","OHI");

KKO(T+1,"OHI")..

K(T+1,"OHI")=E=((1–DELTA/100)**10)*K(T,"OHI")+10*(Y(T,"OHI")–
        C(T,"OHI"));

BCO(T,"OHI").. C(T,"OHI")=L=Y(T,"OHI");

OBJO.. UTILITY=E=SUM(T, STPF(T)*L(T,"OHI")*
        LOG(C(T,"OHI")/L(T,"OHI")));

*EUROPE

YYE(T,"EUROPE").. Y(T,"EUROPE")=E=

(TFP(T,"EUROPE")*(K(T,"EUROPE")**GAMMA)*(L(T,"EUROPE")**(
  1–GAMMA–OW("ALPHA","EUROPE")))*((CA(T,"EUROPE")/
  PHI(T,"EUROPE"))**OW("ALPHA","EUROPE"))–((EPP("MU",
  "EUROPE")+113+700*(CUMCARB(T)/CARBMAX)**EXPCARB)/
  1000)*CA(T,"EUROPE")/PHI(T,"EUROPE"));

KFE(TFIRST,"EUROPE").. K(TFIRST,"EUROPE")=E=OW("K0",
                "EUROPE");

KKE(T+1,"EUROPE")..

K(T+1,"EUROPE")=E=((1–DELTA/100)**10)*K(T,"EUROPE")
  +10*(Y(T,"EUROPE")–C(T,"EUROPE"));

BCE(T,"EUROPE").. C(T,"EUROPE")=L=Y(T,"EUROPE");

OBJE.. UTILITY=E=SUM(T, STPF(T)*L(T,"EUROPE")
                *LOG(C(T,"EUROPE")/L(T,"EUROPE")));

*EE

YYEE(T,"EE").. Y(T,"EE")=E=

(TFP(T,"EE")*(K(T,"EE")**GAMMA)*(L(T,"EE")**(1–GAMMA–
  OW("ALPHA","EE")))*((CA(T,"EE")/PHI(T,"EE"))**OW("ALPHA
  ","EE"))–((EPP("MU","EE")+113+700*(CUMCARB(T)/
  CARBMAX)**EXPCARB)/1000)*CA(T,"EE")/PHI(T,"EE"));

KFEE(TFIRST,"EE").. K(TFIRST,"EE")=E=OW("K0","EE");

KKEE(T+1,"EE")..

K(T+1,"EE")=E=((1–DELTA/100)**10)*K(T,"EE")+10*(Y(T,"EE")–
        C(T,"EE"));

BCEE(T,"EE").. C(T,"EE")=L=Y(T,"EE");

OBJEE.. UTILITY=E=SUM(T, STPF(T)*L(T,"EE")*LOG(C(T,"EE")/
L(T,"EE")));

*MI

YYMI(T,"MI").. Y(T,"MI")=E=

(TFP(T,"MI")*(K(T,"MI")**GAMMA)*(L(T,"MI")**(1−GAMMA−
OW("ALPHA","MI")))*((CA(T,"MI")/PHI(T,"MI"))**OW("ALPHA
","MI"))−((EPP("MU","MI")+113+700*(CUMCARB(T)/
CARBMAX)**EXPCARB)/1000)*CA(T,"MI")/PHI(T,"MI"));

KFMI(TFIRST,"MI").. K(TFIRST,"MI")=E=OW("K0","MI");

KKMI(T+1,"MI")..

K(T+1,"MI")=E=((1−DELTA/100)**10)*K(T,"MI")+10*(Y(T,"MI")−
C(T,"MI"));

BCMI(T,"MI").. C(T,"MI")=L=Y(T,"MI");

OBJMI.. UTILITY=E=SUM(T, STPF(T)*L(T,"MI")*LOG(C(T,"MI")/
L(T,"MI")));

*LMI

YYLM(T,"LMI").. Y(T,"LMI")=E=

(TFP(T,"LMI")*(K(T,"LMI")**GAMMA)*(L(T,"LMI")**(1−GAMMA−
OW("ALPHA","LMI")))*((CA(T,"LMI")/PHI(T,"LMI"))**OW("ALP
HA","LMI"))−((EPP("MU","LMI")+113+700*(CUMCARB(T)/
CARBMAX)**EXPCARB)/1000)*CA(T,"LMI")/PHI(T,"LMI"));

KFLM(TFIRST,"LMI").. K(TFIRST,"LMI")=E=OW("K0","LMI");

KKLM(T+1,"LMI")..

K(T+1,"LMI")=E=((1−DELTA/100)**10)*K(T,"LMI")+10*(Y(T,"LMI")−
C(T,"LMI"));

BCLM(T,"LMI").. C(T,"LMI")=L=Y(T,"LMI");

OBJLM.. UTILITY=E=SUM(T, STPF(T)*L(T,"LMI")
*LOG(C(T,"LMI")/L(T,"LMI")));

*CHINA

YYC(T,"CHINA").. Y(T,"CHINA")=E=

(TFP(T,"CHINA")*(K(T,"CHINA")**GAMMA)*(L(T,"CHINA")**(1−
GAMMA−OW("ALPHA","CHINA")))*((CA(T,"CHINA")/PHI(T,
"CHINA"))**OW("ALPHA","CHINA"))−((EPP("MU","CHINA")
+113+700*(CUMCARB(T)/CARBMAX)**EXPCARB)/1000)*CA(T,"
CHINA")/PHI(T,"CHINA"));

KFC(TFIRST,"CHINA").. K(TFIRST,"CHINA")=E=OW("K0", "CHINA");

KKC(T+1,"CHINA").. K(T+1,"CHINA")=E=((1–DELTA/100)**10) *K(T,"CHINA")+10*(Y(T,"CHINA")–C(T,"CHINA")));

BCC(T,"CHINA").. C(T,"CHINA")=L=Y(T,"CHINA");

OBJC.. UTILITY=E=SUM(T, STPF(T)*L(T,"CHINA")*LOG(C(T, "CHINA")/L(T,"CHINA"))));

*LI

YYLI(T,"LI").. Y(T,"LI")=E=

(TFP(T,"LI")*(K(T,"LI")**GAMMA)*(L(T,"LI")**(1–GAMMA– OW("ALPHA","LI")))*((CA(T,"LI")/PHI(T,"LI"))**OW("ALPHA", "LI"))–((EPP("MU","LI")+113+700*(CUMCARB(T)/ CARBMAX)**EXPCARB)/1000)*CA(T,"LI")/PHI(T,"LI"));

KFLI(TFIRST,"LI").. K(TFIRST,"LI")=E=OW("K0","LI");

KKLI(T+1,"LI")..

K(T+1,"LI")=E=((1–DELTA/100)**10)*K(T,"LI")+10*(Y(T,"LI")– C(T,"LI"));

BCLI(T,"LI").. C(T,"LI")=L=Y(T,"LI");

OBJLI.. UTILITY=E=SUM(T, STPF(T)*L(T,"LI")*LOG(C(T,"LI")/L(T,"LI"))));

MODEL SVUSA   /YYU, KFU, KKU, BCU, OBJU/;

SVUSA.OPTFILE=1;

SOLVE SVUSA MAXIMIZING UTILITY USING NLP;

SOLVE SVUSA MAXIMIZING UTILITY USING NLP;

MODEL SVOHI   /YYO, KFO, KKO, BCO, OBJO/;

SVOHI.OPTFILE=1;

SOLVE SVOHI MAXIMIZING UTILITY USING NLP;

SOLVE SVOHI MAXIMIZING UTILITY USING NLP;

MODEL SVEUROPE   /YYE, KFE, KKE, BCE, OBJE/;

SVEUROPE.OPTFILE=1;

SOLVE SVEUROPE MAXIMIZING UTILITY USING NLP;

SOLVE SVEUROPE MAXIMIZING UTILITY USING NLP;

MODEL SVEE   /YYEE, KFEE, KKEE, BCEE, OBJEE/;

SVEE.OPTFILE=1;

SOLVE SVEE MAXIMIZING UTILITY USING NLP;
SOLVE SVEE MAXIMIZING UTILITY USING NLP;
MODEL SVMI   /YYMI, KFMI, KKMI, BCMI, OBJMI/;
SVMI.OPTFILE=1;
SOLVE SVMI MAXIMIZING UTILITY USING NLP;
SOLVE SVMI MAXIMIZING UTILITY USING NLP;
MODEL SVLMI   /YYLM, KFLM, KKLM, BCLM, OBJLM/;
SVLMI.OPTFILE=1;
SOLVE SVLMI MAXIMIZING UTILITY USING NLP;
SOLVE SVLMI MAXIMIZING UTILITY USING NLP;
MODEL SVCHINA   /YYC, KFC, KKC, BCC, OBJC/;
SVCHINA.OPTFILE=1;
SOLVE SVCHINA MAXIMIZING UTILITY USING NLP;
SOLVE SVCHINA MAXIMIZING UTILITY USING NLP;
MODEL SVLI /YYLI, KFLI, KKLI, BCLI, OBJLI/;
SVLI.OPTFILE=1;
SOLVE SVLI MAXIMIZING UTILITY USING NLP;
SOLVE SVLI MAXIMIZING UTILITY USING NLP;

PARAMETERS

| | | |
|---|---|---|
| SVE(T) | Starting value emissions | (GtC per year) |
| SVM(T) | Starting value concentration | (GtC) |
| SVMU(T) | Starting value upper level conc | (GtC) |
| SVML(T) | Starting value deep ocean conc | (GtC) |
| SVF(T) | Starting value forcing | (W per m squared) |
| SVT(T) | Starting value temperature | (deg C from 1900) |
| SVTL(T) | Starting value deep ocean temperature | (deg C from 1900); |

SVE(T)=SUM(N, CA.L(T,N))+SUM(N, LU(T,N));
SVM('1')=M0;
SVMU('1')=MU0;
SVML('1')=ML0;

```
LOOP (T,
   SVM(T+1)=10*SVE(T)+(TRAA/100)*SVM(T)+(TRUA/100)*
            SVMU(T);
   SVMU(T+1)=(TRAU/100)*SVM(T)+(TRUU/100)*SVMU(T)
     +(TRLU/100)*SVML(T);
   SVML(T+1)=(TRUL/100)*SVMU(T)+(TRLL/100)*SVML(T););

SVF(T)=4.1*LOG(SVM(T)/596.4)/LOG(2)+EXOGFORC(T);

SVT('1')=TE0;

SVTL('1')=TL0;

LOOP (T,
   SVT(T+1)=SVT(T)+SAT*(SVF(T)-(4.1/CS)*SVT(T)-HLAL*(SVT(T)-
     SVTL(T)));
   SVTL(T+1)=SVTL(T)+HGLA*(SVT(T)-SVTL(T)););

***

*** MINOS5.OPT-INCREASE MAXIMUM ITERATIONS FOR
     MINOS SOLVER
BEGIN GAMS/MINOS OPTIONS
Major iterations 6000
END GAMS/MINOS OPTIONS
```

# Appendix E: GAMS Code for DICE-99

```
SETS T       Time periods                                    /1*35/
     TFIRST(T)  First period
     TLAST(T)   Last period
     tearly(T)  First 20 periods
     TLATE(T)   Second 20 periods;

TFIRST(T) = YES$(ORD(T) EQ 1);
TLAST(T) = YES$(ORD(T) EQ CARD(T));
TEARLY(T) = YES$(ORD(T) LE 20);
TLATE(T) = YES$(ORD(T) GE 21);

SCALARS
```

| | | |
|---|---|---|
| A1 | Damage coeff linear term | /−.0045/ |
| A2 | Damage coeff quadratic term | /.0035/ |
| COST10 | Intercept control cost function | /.03/ |
| COST2 | Exponent of control cost function | /2.15/ |
| dmiufunc | Decline in cost of abatement function (pct per decade) | /−8/ |
| decmiu | Change in decline of cost function (pct per year) | /.5/ |
| DK | Depreciation rate on capital (pct per year) | /10/ |
| GAMA | Capital elasticity in production function | /.30/ |
| K0 | 1990 value capital trillion 1990 US dollars | /47/ |
| LU0 | Initial land use emissions (GtC per year) | /1.128/ |
| SIG0 | CO2-equivalent emissions-GNP ratio | /.274/ |

| GSIGMA | Growth of sigma (pct per decade) | /−15.8854/ |
| desig | Decline rate of decarbonization (pct per decade) | /2.358711/ |
| desig2 | Quadratic term in decarbonization | /−.00085/ |
| WIEL | World industrial emissions limit (GtC per year) | /5.67/ |
| LL0 | 1990 world population (millions) | /5632.7/ |
| GL0 | Initial rowth rate of population (pct per decade) | /15.7/ |
| DLAB | Decline rate of pop growth (pct per decade) | /22.2/ |
| A0 | Initial level of total factor productivity | /.01685/ |
| GA0 | Initial growth rate for technology (pct per decade) | /3.8/ |
| DELA | Decline rate of technol. change per decade | /.000001/ |
| MAT1990 | Concentration in atmosphere 1990 (b.t.c.) | /735/ |
| MU1990 | Concentration in upper strata 1990 (b.t.c) | /781/ |
| ML1990 | Concentration in lower strata 1990 (b.t.c) | /19230/ |
| b11 | Carbon cycle transition matrix (pct per decade) | /66.616/ |
| b12 | Carbon cycle transition matrix | /33.384/ |
| b21 | Carbon cycle transition matrix | /27.607/ |
| b22 | Carbon cycle transition matrix | /60.897/ |
| b23 | Carbon cycle transition matrix | /11.496/ |
| b32 | Carbon cycle transition matrix | /0.422/ |
| b33 | Carbon cycle transition matrix | /99.578/ |
| TL0 | 1985 lower strat. temp change (C) from 1900 | /.06/ |
| T0 | 1985 atmospheric temp change (C)from 1900 | /.43/ |
| C1 | Climate-equation coefficient for upper level | /.226/ |
| CS | Eq temp increase for CO2 doubling (C) | /2.9078/ |
| C3 | Transfer coeffic. upper to lower stratum | /.440/ |
| C4 | Transfer coeffic for lower level | /.02/ |
| SRTP | Initial rate of social time preference (pct per year) | /3/ |

| DR | Decline rate of social time preference (pct per year) | /.25719/ |
|---|---|---|
| coefopt1 | Scaling coefficient in the objective function | /333.51/ |
| coefopt2 | Scaling coefficient in the objective function | /622.78/; |

PARAMETERS

| cost1(t) | cost function for abatement |
|---|---|
| gcost1(t) | |
| ETREE(T) | Emissions from deforestation |
| GSIG(T) | Cumulative improvement of energy efficiency |
| SIGMA(T) | CO2-equivalent-emissions output ratio |
| WEL(T) | World total emissions limit (GtC) |
| L(T) | Level of population and labor |
| GL(T) | Growth rate of labor 0 to T |
| AL(T) | Level of total factor productivity |
| GA(T) | Growth rate of productivity from 0 to T |
| FORCOTH(T) | Exogenous forcing for other greenhouse gases |
| R(T) | Instantaeous rate of social time preference |
| RR(T) | Average utility social discount rate; |

gcost1(T)=(dmiufunc/100)*EXP(−(decmiu/100)*10*(ORD(T)−1));

cost1("1")=cost10;

LOOP(T,

cost1(T+1)=cost1(T)/((1+gcost1(T+1))););

ETREE(T) = LU0*(1−0.1)**(ord(T)−1);

gsig(T)=(gsigma/100)*EXP(−(desig/100)*10*(ORD(T)−1) − desig2*10*
  ((ord(t)−1)**2));

sigma("1")=sig0;

LOOP(T,

sigma(T+1)=(sigma(T)/((1−gsig(T+1)))););

WEL(T)=WIEL+ETREE(T);

GL(T) = (GL0/DLAB)*(1−exp(−(DLAB/100)*(ord(t)−1)));

L(T)=LL0*exp(GL(t));

ga(T)=(ga0/100)*EXP(−(dela/100)*10*(ORD(T)−1));

al("1") = a0;

LOOP(T,

al(T+1)=al(T)/((1−ga(T))););

FORCOTH(T)=(−0.1965+(ORD(T)−1)*0.13465)$

   (ORD(T) LT 12) + 1.15$(ORD(T) GE 12);

R(T)=(srtp/100)*EXP(−(DR/100)*10*(ORD(T)−1));

RR("1")=1;

LOOP(T,

RR(T+1)=RR(T)/((1+R(T))**10););

VARIABLES

| | |
|---|---|
| Y(T) | Output |
| I(T) | Investment trill US dollars |
| K(T) | Capital stock trill US dollars |
| E(T) | CO2-equivalent emissions bill t |
| MIU(T) | Emission control rate GHGs |
| MAT(T) | Carbon concentration in atmosphere (b.t.c.) |
| MU(T) | Carbon concentration in shallow oceans (b.t.c.) |
| ML(T) | Carbon concentration in lower oceans (b.t.c.) |
| TE(T) | Temperature of atmosphere (C) |
| FORC(T) | Radiative forcing (W per m2) |
| TL(T) | Temperature of lower ocean (C) |
| C(T) | Consumption trill US dollars |

UTILITY;

POSITIVE VARIABLES MIU, TE, E, Mat, mu, ml, Y, C, K, I;

EQUATIONS

| | |
|---|---|
| YY(T) | Output equation |
| CC(T) | Consumption equation |
| KK(T) | Capital balance equation |
| KK0(T) | Initial condition for K |
| KC(T) | Terminal condition for K |
| EE(T) | Emissions process |
| MMAT0(T) | Starting atmospheric concentration |

MMAT(T)      Atmospheric concentration equation

MMU0(T)      Initial shallow ocean concentration

MMU(T)       Shallow ocean concentration

MML0(T)      Initial lower ocean concentration

MML(T)       Lower ocean concentration

TTE(T)       Temperature-climate equation for atmosphere

TTE0(T)      Initial condition for atmospheric temperature

FORCE(T)     Radiative forcing equation

TLE(T)       Temperature-climate equation for lower oceans

TLE0(T)      Initial condition for lower ocean

UTIL         Objective function;

** Equations of the model

KK(T)..   K(T+1) =L= (1−(DK/100))**10 *K(T)+10*I(T);

KK0(TFIRST)..   K(TFIRST) =E= K0;

KC(TLAST)..   .02*K(TLAST) =L= I(TLAST);

EE(T)..   E(T)=G=10*SIGMA(T)*(1−(MIU(T)/100))*AL(T)*L(T)**(1−
   GAMA)*K(T)**GAMA + ETREE(T);

FORCE(T)..   FORC(T) =E= 4.1*((log(Mat(T)/596.4)/log(2)))
   +FORCOTH(T);

MMAT0(TFIRST)..   MAT(TFIRST) =E= MAT1990;

MMU0(TFIRST)..   MU(TFIRST) =E= MU1990;

MML0(TFIRST)..   ML(TFIRST) =E= ML1990;

MMAT(T+1)..   MAT(T+1) =E= MAT(T)*(b11/100)+E(T)
   +MU(T)*(b21/100);

MML(T+1)..   ML(T+1) =E= ML(T)*(b33/100)+(b23/100)*MU(T);

MMU(T+1)..   MU(T+1) =E= MAT(T)*(b12/100)+MU(T)*(b22/100)
   +ML(T)*(b32/100);

TTE0(TFIRST)..   TE(TFIRST) =E= T0;

TTE(T+1)..   TE(T+1) =E= TE(t)+C1*(FORC(t)−(4.1/CS)*TE(t)−
   C3*(TE(t)−TL(t)));

TLE0(TFIRST)..   TL(TFIRST) =E= TL0;

TLE(T+1)..   TL(T+1) =E= TL(T)+C4*(TE(T)−TL(T));

YY(T)..   Y(T) =E= AL(T)*L(T)**(1−GAMA)*K(T)**GAMA*(1−
   cost1(t)*((MIU(T)/100)**cost2))/(1+a1*TE(T)+ a2*TE(T)**2);

CC(T)..   C(T) =E= Y(T)–I(T);

UTIL..   UTILITY =E= SUM(T, 10 *RR(T)*L(T)*LOG(C(T)/L(T)))/
  coefopt1)+coefopt2;

** Upper and Lower Bounds: General conditions imposed for
  stability

MIU.up(T) = 1.0;

MIU.lo(T)  = 0.000001;

K.lo(T)     = 1;

TE.up(t)    = 12;

MAT.lo(T) = 10;

MU.lo(t)    = 100;

ML.lo(t)    = 1000;

C.lo(T)     = 2;

** Emissions control policy. Current setting is for optimal policy.
** Reinstate equation "Miu.fx(t) = .0" for no-control run.
MIU.fx(t)= 0;
** Solution options

option iterlim = 99900;

option reslim = 99999;

option solprint = on;

option limrow = 0;

option limcol = 0;

model CO2 /all/;
solve CO2 maximizing UTILITY using nlp;
** Display of results
display y.l, e.l, mat.l, te.l;

Parameters

Year(t)      Date

Indem(t)   Industrial emissions (b.t.c. per year)

Wem(t)     Total emissions (b.t.c. per year)

S(t)          Savings rate (pct);

Year(t)    = 1995 +10*(ord(t)–1);

Indem(t) = e.l(t)–etree(t);

```
Wem(t)   = e.l(t);
S(t) = 100*i.l(t)/y.l(t);

display s;

Parameters
Tax(t)        Carbon tax ($ per ton)
damtax(t)   Concentration tax ($ per ton)
dam(t)        Damages
cost(t)        Abatement costs;

tax(t)       = -1*ee.m(t)*1000/(kk.m(t));
damtax(t) = -1*mmat.m(t)*1000/kk.m(t);
dam(t)      = y.l(t)*(1-1/(1+a1*te.l(t)+ a2*te.l(t)**2));
cost(t)      = y.l(t)*(cost1(t)*(miu.l(t)**cost2));

File d99oute;
D99oute.pc=5;
D99oute.pw=250;
Put d99oute;
Put / "base (no control) run";
Put / "year";
Loop (tearly, put year(tearly)::0);
Put / "output";
Loop (tearly, put y.l(tearly)::3);
Put / "indem";
Loop (tearly, put indem(tearly)::4);
Put / "sigma";
Loop (tearly, put sigma(tearly)::4);
Put / "temp";
Loop (tearly, put te.l(tearly)::3);
Put / "conc";
Loop (tearly, put mat.l(tearly)::3);
Put / "ctax";
Loop (tearly, put tax(tearly)::2);
Put / "discrate";
```

```
Loop (tearly, put rr(tearly)::5);
Put / "prod";
Loop (tearly, put al(tearly)::3);
Put / "exogforc";
Loop (tearly, put forcoth(tearly)::3);
Put / "pop";
Loop (tearly, put l(tearly)::3);
Put / "etree";
Loop (tearly, put etree(tearly)::4);
Put / "margy";
Loop (tearly, put yy.m(tearly)::3);
Put / "margc";
Loop (tearly, put cc.m(tearly)::3);
Put / "miu";
Loop (tearly, put miu.l(tearly)::3);
Put / "total emissions";
Loop (tearly, put wem(tearly)::3);
Put / "damages";
Loop (tearly, put dam(tearly)::5);
Put / "abatement cost";
Loop (tearly, put cost(tearly)::5);
Put /"objective function";
Put utility.l::3;

File d99outL;
D99outL.pc=5;
D99outL.pw=250;
Put d99outL;
Put / "base (no control) run";
Put / "year";
Loop (tlate, put year(tlate)::0);
Put / "output";
Loop (tlate, put y.l(tlate)::3);
```

Put / "indem";
Loop (tlate, put indem(tlate)::4);
Put / "sigma";
Loop (tlate, put sigma(tlate)::4);
Put / "temp";
Loop (tlate, put te.l(tlate)::3);
Put / "conc";
Loop (tlate, put mat.l(tlate)::3);
Put / "ctax";
Loop (tlate, put tax(tlate)::2);
Put / "discrate";
Loop (tlate, put rr(tlate)::5);
Put / "prod";
Loop (tlate, put al(tlate)::3);
Put / "exogforc";
Loop (tlate, put forcoth(tlate)::3);
Put / "pop";
Loop (tlate, put l(tlate)::3);
Put / "etree";
Loop (tlate, put etree(tlate)::4);
Put / "margy";
Loop (tlate, put yy.m(tlate)::3);
Put / "margc";
Loop (tlate, put cc.m(tlate)::3);
Put / "miu";
Loop (tlate, put miu.l(tlate)::3);
Put / "total emissions";
Loop (tlate, put wem(tlate)::3);
Put / "damages";
Loop (tlate, put dam(tlate)::5);
Put / "abatement cost";
Loop (tlate, put cost(tlate)::5);

# References

Broecker, Wallace S. 1997. "Thermohaline Circulation, the Achilles Heel of Our Climate System: Will Man-Made CO$_2$ Upset the Current Balance." *Science* 278 (November 28): 1582–1588.

Cline, William. 1992a. *The Economics of Global Warming*. Washington, D.C.: Institute of International Economics.

Dansgaard, W. et al. 1993. "Evidence for General Instability of Past Climate from a 250-Kyr Ice-Core Record." *Nature* (July 15): 218–220.

Darwin, Roy, Marinos Tsigas, Jan Lewandrowski, and Anton Raneses. 1995. *World Agriculture and Climate Change*. U. S. Department of Agriculture, Agricultural Economic Report No. 703, June.

Dinar, Ariel, Robert Mendelsohn, Robert Evenson, Jyoti Parikh, Apurva Sanghi, Kavi Kumar, James McKinsey, and Stephen Lonergan. 1998. *Measuring the Impact of Climate Change on Indian Agriculture*. World Bank Technical Paper No. 402.

Eisner, Robert. 1989. *The Total Incomes System of Accounts*. Chicago: University of Chicago Press.

Energy Information Agency, U.S. Department of Energy. 1996. *Emissions of Greenhouse Gases in the United States, 1996*. Washington, D.C.: GPO.

Fankhauser, Samuel. 1995. *Valuing Climate Change: The Economics of the Greenhouse Effect*. London: Earthscan.

IPCC. 1990. Intergovernmental Panel on Climate Change, *Climate Change: The IPCC Scientific Assessment*, J. T. Houghton, G. J. Jenkins, and J. J. Ephraums, eds. U.K. and NY: Cambridge University Press.

IPCC. 1995. Intergovernmental Panel on Climate Change, *Climate Change 1994: Radiative Forcing of Climate Change and An Evaluation of the IPCC IS92 Emissions Scenarios*. Cambridge, U.K.: Cambridge University Press.

IPCC. 1996a. Intergovernmental Panel on Climate Change, *Climate Change 1995: The Science of Climate Change. The Contribution of Working Group I to the Second Assessment Report of the Intergovernmental Panel on Climate Change*, J. P. Houghton, L. G. Meira Filho, B. A. Callendar, A. Kattenberg, and K. Maskell, eds. Cambridge, U.K.: Cambridge University Press.

IPCC. 1996b. Intergovernmental Panel on Climate Change, *Climate Change 1995: Impacts, Adaptation, and Mitigation of Climate Change: Scientific-Technical Analysis. The Contribution of Working Group II to the Second Assessment Report of the Intergovernmental Panel on Climate Change*, R. T. Watson, M. C. Zinyowera, R. H. Moss, eds. Cambridge, U.K.: Cambridge University Press.

IPCC. 1996c. Intergovernmental Panel on Climate Change *Climate Change 1995: Economic and Cross-Cutting Issues. The Contribution of Working Group III to the Second Assessment Report of the Intergovernmental Panel on Climate Change*, J. P. Bruce, H. Lee, and E. F. Haites, eds. Cambridge, U.K.: Cambridge University Press.

Kolstad, Charles D. 1998. "Integrated Assessment Modeling of Climate Change." In Nordhaus 1998b.

Koopmans, Tjalling. 1967. "Objectives, Constraints, and Outcomes in Optimal Growth Models." *Econometrica* 35: 1–15.

Maddison, Angus. 1995. *Monitoring the World Economy 1820–1992*. Paris: OECD Development Centre.

———. 1998a. *Chinese Economic Performance in the Long Run*. Paris and Washington, D.C.: OECD Development Centre.

———. 1998b. "Economic Growth Since 1500 A.D.: Problems of Measurement, Interpretation, and Explanation." Twelfth Annual Kuznets Lectures. Yale University Department of Economics, November 4–6.

Mendelsohn, Robert, and James E. Neumann. 1999. *The Impact of Climate Change on the United States Economy*. Cambridge, U.K.: Cambridge University Press.

Murray, Christopher J. L., and Alan D. Lopez, eds. 1996. *The Global Burden of Disease*. Harvard School of Public Health. Cambridge, MA: Harvard University Press.

Nakicenovic, Nebojsa, Arnulf Grubler, and Alan McDonald, eds. 1998. *Global Energy Perspectives*. Cambridge, U.K.: Cambridge University Press.

National Academy of Sciences. 1992. Committee on Science, Engineering, and Public Policy, *Policy Implications of Greenhouse Warming: Mitigation, Adaptation, and the Science Base*. Washington, D.C.: National Academy Press.

Negishi, T. 1960. "Welfare Economics and the Existence of an Equilibrium for a Competitive Economy." *Metroeconomica* 12: 92–97.

Nordhaus, William D. 1989. "The Economics of the Greenhouse Effect." *Paper presented at the International Energy Workshop*. Laxenburg, Austria, June.

———. 1990a. "Slowing the Greenhouse Express: The Economics of Greenhouse Warming." In Henry Aaron, ed. *Setting National Priorities*. Washington, D.C.: Brookings Institution.

———. 1991a. "To Slow or Not to Slow: The Economics of the Greenhouse Effect." *The Economic Journal* 101 (July): 920–937.

———. 1994a. "Expert Opinion on Climatic Change." *American Scientist* 82 (January–February): 45–51.

———. 1994b. *Managing the Global Commons: The Economics of Climate Change*. Cambridge, MA: MIT Press.

——. 1998a. "Discounting and Public Policies That Affect the Distant Future." In Portney and Weyant. 1999.

——. ed. 1998b. *Economic and Policy Issues in Climate Change*. Washington, D.C.: Resources for the Future Press.

——. 1998c. "Impact of Climate and Climate Change on Non-Market Time Use." Yale University, processed, September 30.

Nordhaus, William D., and Joseph Boyer. 1999. "Requiem for Kyoto: An Economic Analysis of the Kyoto Protocol." In Weyant. 1999.

Nordhaus, William D., and James Tobin. 1972. "Is Growth Obsolete?" In *Economic Growth*, Fiftieth Anniversary Colloquium V, National Bureau of Economic Research. New York: Columbia University Press.

Nordhaus, William D., and Zili Yang. 1996. "A Regional Dynamic General-Equilibrium Model of Alternative Climate-Change Strategies." *American Economic Review* 86, no. 4 (September): 741–765.

Portney, Paul, and John Weyant, eds. 1999. *Discounting and Intergenerational Equity*. Washington, D.C.: Resources for the Future.

Ramsey, Frank P. 1928. "A Mathematical Theory of Saving." *The Economic Journal* (December): 543–559.

Robinson, John W., and Geoffrey Godbey. 1997. *Time for Life: The Surprising Ways Americans Use Their Time*. University Park, PA: Pennsylvania University Press.

Rogner, Hans Holger. 1997. "An Assessment of World Hydrocarbon Resources." *Annual Review of Energy and the Environment* 22: 217–262.

Rosenthal, D. H., H. K. Gruenspecht, and E. A. Moran. 1994. *Effects of Global Warming on Energy Use for Space Heating and Cooling in the United States*. Mimeo. Washington, D.C.: U.S. Department of Energy.

Sanghi, Apurva, Robert Mendelsohn, and Ariel Dinar. 1998. "The Climate Sensitivity of Indian Agriculture." In Dinar et al. 1998.

Schimmelpfennig, David. 1996. *Agricultural Adaptation to Climate Change*. U. S. Department of Agriculture, Agricultural Economic Report No. 740, June.

Schlesinger, Michael E., and Xingjian Jiang. 1990. "Simple Model Representation of Atmosphere-Ocean GCMs and Estimation of the Timescale of $CO_2$-Induced Climate Change." *Journal of Climate* (December): 12–15.

Schneider, Stephen H., and Starley L. Thompson. 1981. "Atmospheric $CO_2$ and Climate: Importance of the Transient Response." *Journal of Geophysical Research* 86, no. C4 (April 20): 3135–3147.

Schultz, Peter A., and James F. Kasting. 1997. "Optimal Reductions in $CO_2$ Emissions." *Energy Policy* 25, no. 5: 491–500.

Solow, Robert M. 1970. *Growth Theory: An Exposition*. New York: Oxford University Press.

Statistical Abstract. 1997. *Statistical Abstract of the United States*. Washington, D.C.: GPO.

Stocker, T. F., and A. Schmitter. 1997. "Rate of Global Warming Determines the Stability of the Ocean-Atmosphere System." *Nature* 388: 862–865.

Stouffer, R. J., S. Manabe, and K. Bryan. 1989. "Interhemispheric Asymmetry in Climate Response to a Gradual Increase of Atmospheric $CO_2$." *Nature* 342 (December 7): 660–662.

Taylor, K. C., et al. 1993. "The 'Flickering Switch' of Late Pleistocene Climate Change." *Nature* 361 (February 4): 432–436.

Tol, R. S. J. 1995. "The Damage Costs of Climate Change: Towards More Comprehensive Calculations." *Environmental and Resource Economics* 5: 353–374.

Tolley, George S., Donald Scott Kenkel, and Robert Fabian, eds. 1994. *Valuing Health for Policy*. Chicago, Illinois: University of Chicago Press.

Toth, Ferenc et al. 1998. "Kyoto and the Long-term Climate Stabilization." Working paper, PIK (Potsdam Institute for Climate Impact Research).

Weyant, John P., ed. 1999. *The Costs of the Kyoto Protocol: A Multi-Model Evaluation*. Special issue of *The Energy Journal*.

Wigley, T. M. L. 1998. "The Kyoto Protocol: $CO_2$, $CH_4$, and Climate Implications." *Geophysical Research Letters* 25, no. 13 (July 1): 2285–2288.

Wigley, T. M. L., M. Solomon, and S. C. B. Raper. 1994. "Model for the Assessment of Greenhouse-Gas Induced Climate Change." Version 1.2. Unversity of East Anglia, U.K.: Climate Research Unit.

Yohe, Gary W., and Michael E. Schlesinger. 1998. "Sea-Level Change: The Expected Economic Cost of Protection or Abandonment in the United States." *Climatic Change* 38: 337–342.

## Other Useful Sources

Amano, Akihiro. 1992. "Economic Costs of Reducing $CO_2$ Emissions: A Study of Modeling Experience in Japan." In Kaya et al. 1993.

Ausubel, Jesse H. 1993. "Mitigation and Adaptations for Climate Change: Answers and Questions." In Kaya et al. 1993.

Ausubel, Jesse H., and William D. Nordhaus. 1983. "A Review of Estimates of Future Carbon Dioxide Emissions." In National Research Council: 153–185.

Boden, Thomas A., Paul Kanciruk, and Michael P. Farrell eds. 1990. *Trends '90: A Compendium of Data on Global Change*. Carbon Dioxide Information Analysis Center, ORNL/CDIAC-36, August. Oak Ridge, TE: Oak Ridge National Laboratory.

Broecker, W. S., and T. H. Peng. 1982. "Tracers in the Sea." Palisades, NY: Eldigio Press, Lamont-Doherty Geological Observatory.

Brooke, Anthony, David Kendrick, and Alexander Meeraus. 1988. *GAMS: A User's Guide*. Redwood City, CA: The Scientific Press.

Cline, William. 1991. "The Economics of the Greenhouse Effect." *The Economic Journal* 101 (July): 920–937.

———. 1992b. "Discounting." Paper presented to the International Workshop on Costs, Impacts, and Possible Benefits of $CO_2$ Mitigation. September. Laxenburg, Austria: IIASA.

Coolfont Workshop. 1989. *Climate Impact Response Functions: Report of a Workshop Held at Coolfont, West Virginia.* September 11–14. Washington, DC: National Climate Program Office.

Dean, Andrew, and Peter Hoeller. 1992. *Costs of Reducing $CO_2$ Emissions: Evidence from Six Global Models.* OCDE/GD(92)140. Organisation for Economic Cooperation and Development, processed, Paris.

Dubin, Jeffrey A. 1992. "Market Barriers to Conservation: Are Implicit Discount Rates Too High?" In Matthew G. Nagler, ed. *The Economics of Energy Conservation.* Proceedings of a POWER Conference, Berkeley, CA.

EC. 1992a. *The Climate Challenge: Economic Aspects of the Community's Strategy for Limiting $CO_2$ Emissions.* Commission of the European Communities. No. 51, ECSC-EEC-EAEC. Brussels. May.

EC. 1992b. *The Economics of Limiting $CO_2$ Emissions.* Commission of the European Communities, Special edition no. 1,ECSC-EEC-EAEC. Brussels.

Edmonds, J. A., and J. M. Reilly. 1983. "Global Energy and $CO_2$ to the Year 2050." *The Energy Journal* 4: 21–47.

Edmonds, J. A., J. M. Reilly, R. H. Gardner, and A. Brenkert. 1986. *Uncertainty in Future Energy Use and Fossil Fuel $CO_2$ Emissions 1975 to 2075.* DOE/NBB-0081. Department of Energy, Washington, D.C., December.

Eizenstat, Stuart. 1998. *Statement.* Testimony before Subcommittee on Energy and Power, U.S. Congress, House Commerce Committee, March 4.

EPA. 1989. *The Potential Effects of Global Climate Change on the United States: Report to Congress.* EPA-230-05-89-050. U. S. Environmental Protection Agency, December.

EPA. 1990. Lashof Daniel A., and Dennis A. Tirpak, eds. *Policy Options for Stabilizing Global Climate.* New York: Hemisphere Pub. Corp.

Gaskins, Darius W., and John P. Weyant. 1993. "EMF-12: Modeling Comparisons of the Costs of Reducing $CO_2$ Emissions." *American Economic Review* (May): 318–323.

Gordon, Robert, Tjalling Koopmans, William Nordhaus, and Brian Skinner. 1988. *Toward a New Iron Age?* Cambridge, MA: Harvard University Press.

GRIP. 1993. Greenland Ice-Core Project (GRIP) Members. "Climate Instability during the Last Interglacial Period Recorded in the GRIP Ice Core." *Nature* (July 15): 203–208.

Grossman, Sanford J., and Robert J. Shiller. 1981. "The Determinants of the Variability of Stock Market Prices." *American Economic Review* 71 (May): 222–227.

Hammitt, James K., Robert J. Lempert, and Michael E. Schlesinger. 1992. "A Sequential-Decision Strategy for Abating Climate Change." *Nature* 357 (May 28): 315–318.

Henrion, M., and B. Fischoff. 1986. "Assessing Uncertainty in Physical Constants." *American Journal of Physics* 54, no. 9 (September): 791–798.

Hoeller, Peter, Andrew Dean, and Masahiro Hayafumi. 1992. *New Issues, New Results: The OECD's Second Survey of the Macroeconomic Costs of Reducing $CO_2$ Emissions.* OCDE/GD(92)141. Organisation for Economic Cooperation and Development, processed, Paris.

Ibbotson, Roger G., and Gary P. Brinson. 1987. *Investment Markets.* New York: McGraw-Hill.

Jones, P. D., T. M. L. Wigley, and P. B. Wright. 1990. *Global and Hemispheric Annual Temperature Variations Between 1861 and 1988.* NDP-022/R1. Oak Ridge, TE: Carbon Dioxide Information Center, Oak Ridge National Laboratory.

Jorgenson, Dale W., and Peter J. Wilcoxen. 1990. "The Cost of Controlling U. S. Carbon Dioxide Emissions." Paper presented at a Workshop on Economic/Energy/Environmental Modeling for Climate Policy Analysis, Washington, D.C., October.

Jorgenson, Dale W., and Peter J. Wilcoxen. 1991. "Reducing U. S. Carbon Dioxide Emissions: The Cost of Different Goals." In John R. Moroney, ed. *Energy, Growth, and the Environment:* 125–128. Greenwich, CT: JAI Press.

Kasting, James F., and James C. G. Walker. 1992. "The Geochemical Carbon Cycle and the Uptake of Fossil Fuel $CO_2$." In B. G. Levi, D. Hafemeister, and R. Scribner, eds. *Global Warming: Physics and Facts.* AIP Conference Proceedings 247: 175–200. New York: American Institute of Physics.

Kaya, Y., N. Nakicenovic, W. D. Nordhaus, and F. L. Toth. 1993. *Costs, Impacts, and Benefits of $CO_2$ Mitigation.* CP-93-2. International Institute for Systems Analysis. Laxenburg, Austria.

Kolstad, Charles D. 1993. "Looking vs. Leaping: The Timing of $CO_2$ Control in the Face of Uncertainty and Learning." In Kaya et al. 1993.

Lashof, Daniel A., and Dilip R. Ahuja. 1990. "Relative Contributions of Greenhouse Gas Emissions to Global Warming." *Nature* 344 (April 5): 529–531.

Levhari, D., and T. N. Srinivasan. 1969. "Optimal Savings Under Uncertainty." *Review of Economic Studies* 36 (April): 153–163.

Lind, Robert C., ed. 1982. *Discounting for Time and Risk in Energy Policy.* Washington, D.C.: Resources for the Future.

Lindzen, Richard. 1992. "Global Warming." *Regulation*, Summer.

Luce, R. Duncan, and Howard Raiffa. 1958. *Games and Decisions: Introduction and Critical Surveys.* New York: Wiley and Sons.

Machta, Lester. 1972. "The Role of the Oceans and Biosphere in the Carbon Dioxide Cycle." *Nobel Symposium* 20: 121–145.

Maier-Reimer, E., and K. Hasselmann. 1987. "Transport and Storage of Carbon Dioxide in the Ocean, and an Organic Ocean-circulation Carbon Cycle Model." *Climate Dynamics* 2: 63–90.

Manabe, S., and R. J. Stouffer. 1988. "Two Stable Equilibria of a Coupled Ocean-Atmosphere Model." *Journal of Climate* 1: 841–866.

Manabe, S., R. J. Stouffer, M. J. Spelman, and K. Bryan. 1991. "Transient Response of a Coupled Ocean-Atmospheric Model to Gradual Changes of Atmospheric $CO_2$, Part I: Annual Mean Response." *Journal of Climate* 4: 785–818.

Manabe, S., and R. J. Stouffer. 1993. "Century-Scale Effects of Increased Atmospheric $CO_2$ on the Ocean-Atmospheric System." *Nature* 364 (July 15): 215–218.

Manne, Alan S., and Richard G. Richels. 1990a. "$CO_2$ Emission Limits: An Economic Cost Analysis for the USA." *The Energy Journal* 11, no. 2 (April): 51–74.

———. 1990b. "Estimating the Energy Conservation Parameter." processed, November.

———. 1992. *Buying Greenhouse Insurance: The Economic Costs of $CO_2$ Emission Limits.* Cambridge, MA: MIT Press.

Mendelsohn, Robert. 1998. "Overview of Damage Estimates." Paper presented at the EMF Workshop.

Mendelsohn, Robert, William Nordhaus, and Dai Gee Shaw. 1993. "The Impact of Climate on Agriculture: A Ricardian Approach." In Kaya et al. 1993.

Mendelsohn, Robert, William D. Nordhaus, and DaiGee Shaw. 1994. "The Impact of Global Warming on Agriculture: A Ricardian Approach." *American Economic Review* 84, no. 4 (September): 753–771.

Mendelsohn, Robert, William D. Nordhaus, and DaiGee Shaw. 1996. "Climate Impacts on Aggregate Farm Values: Accounting for Adaptation." *Agriculture and Forest Meteorology* 80.

Morgan, M. G., and M. Henrion. 1990. *Uncertainty: A Guide to Dealing with Uncertainty in Quantitative Risk and Policy Analysis.* New York: Cambridge University Press.

Nakicenovic, Nebojsa, William D. Nordhaus, Richard Richels, and Ferenc Toth. 1994. *Integrative Assessment of Mitigation, Impacts, and Adaptation to Climate Change.* CP-94-9, IIASA. Laxenburg, Austria. Also published as a special issue of *Energy Policy* 23, no. 4/5 (April 1995).

Nakicenovic, Nebojsa, William D. Nordhaus, Richard Richels, and Ferenc Toth, eds. 1996. *Climate Change: Integrating Science, Economics, and Policy.* CP-96-1, IIASA. Laxenburg, Austria.

National Academy of Sciences. 1979. *Stratospheric Ozone Depletion by Halocarbons: Chemistry and Transport.* Report of a committee of the National Research Council. Washington, D.C.: National Research Council.

———. 1983. *Changing Climate.* Washington, D.C.: National Academy Press.

———. 1991. *Policy Implications of Greenhouse Warming.* Washington, D.C.: National Academy Press.

National Research Council. 1978. *International Perspectives on the Study of Climate and Society.* Washington, D.C.: National Academy Press.

———. 1979. *Carbon Dioxide and Climate: A Scientific Assessment.* Washington, D.C.: National Academy Press.

———. 1983. *Changing Climate.* Washington, D.C.: National Academy Press.

Nordhaus,William D. 1977. "Economic Growth and Climate: The Case of Carbon Dioxide." *The American Economic Review.* May.

———. 1979. *The Efficient Use of Energy Resources.* New Haven, CT: Yale University Press.

———. 1982."How Fast Should We Graze the Global Commons?" *American Economic Review* 72, no. 2 (May).

———. 1990b. "A General Equilibrium Model of Policies to Slow Global Warming." In David Wood, ed. *Economic Models of Energy and Environment, Proceedings of a Workshop.* Washington, D.C. October.

———. 1991b. "The Cost of Slowing Climate Change: A Survey." *The Energy Journal* 12, no. 1 (Spring): 37–65.

———. 1991c. "Economic Approaches to Greenhouse Warming." In Rudiger Dornbusch and James M. Poterba. *Global Warming: Economic Policy Response*: 33–66. Cambridge, MA: MIT Press.

———. 1991d. "Economic Policy in the Face of Global Warming." In J. W. Tester, D. O. Wood, and N. A. Ferrari. *Energy and the Environment in the 21st Century.* Cambridge, MA: MIT Press.

———. 1992a. "An Optimal Transition Path for Controlling Greenhouse Gases." *Science* 258 (November 20): 1315–1319.

———. 1992b. "How Much Should We Invest to Slow Climate Change?" In Herbert Giersch, ed. *Economic Growth and Evolution.* Berlin: Verlag-Springer.

———. 1993a. "Survey on the Uncertainties Associated with Future Climate Change." Yale University, processed.

———. 1993b. "The Economics of Greenhouse Warming: What are the Issues?" In *Costs, Impacts, and Benefits of CO$_2$ Mitigation.* Y. Kaya, N. Nakicenovic, W. Nordhaus, and F. Toth, eds. CP-93–1993, IIASA, Laxenburg, Austria.

———. 1993c. "Rolling the 'DICE': An Optimal Transition Path for Controlling Greenhouse Gases." *Resource and Energy Economics* 15: 27–50.

———. 1993d. "Optimal Greenhouse-Gas Reductions and Tax Policy in the "DICE" Model." *American Economic Review* 83 (May): 313–317.

———. 1994c. "Climate and Economic Development: Climates Past and Climate Change Future." *Proceedings of the World Bank Annual Conference on Development Economics, 1993–1994*: 355–376.

———. 1995. "The Ghosts of Climates Past and the Specters of Climate Change Future." *Energy Policy* 23, no. 4/5: 269–282.

———. 1996. "Climate Amenities and Global Warming." In *Climate Change: Integrating Science, Economics, and Policy.* N. Nakicenovic, W. Nordhaus, R. Richels, and F. Toth, eds. IIASA, CP-96-1: 3–45.

———. 1997. "Discounting in Economics and Climate Change." *Climatic Change* 37: 315–328.

Nordhaus, William D., and David Popp. 1997. "What Is the Value of Scientific Knowledge? An Application to Global Warming Using the PRICE Model." *The Energy Journal* 18, no. 1: 1–45.

Nordhaus, William D., and Gary Yohe. 1983. "Future Carbon Dioxide Emissions from Fossil Fuels." In National Research Council.

Nuclear Regulatory Commission. 1975. *Reactor Safety Study: An Assessment of Accident Risks in U. S. Commercial Nuclear Power Plants.* NUREG-75/014 WASH-1400. Washington, D.C.

Parikh, Jhodi. 1992. "Emissions Limitations: The View From the South." *Nature* 360 (December): 507–508.

Peck, Stephen C., and Thomas J. Teisberg. 1992. "CETA: A Model for Carbon Emissions Trajectory Assessment." *The Energy Journal* 13, no. 1: 55–77.

Psacharopoulos, George. 1985. "Returns to Education: A Further International Update and Implications." *Journal of Human Resources* 20 (Fall): 583–604.

Ravelle, Roger R., and Paul E. Waggoner. 1983. "Effects of a Carbon Dioxide-Induced Climatic Change on Water Supplied in the Western United States." In National Research Council: 419–432.

Samuelson, Paul A. 1949. "The Market Mechanism and Maximization, I, II, and III." Mimeo. Santa Monica, CA: Rand Corporation.

Savage, L. J. 1954. *The Foundations of Statistics*, New York: Wiley and Sons.

Schelling, Thomas C. 1983. "Climatic Change: Implications for Welfare and Policy." In National Research Council: 449–482.

Schmalensee, Richard. 1993. "Comparing Greenhouse Gases for Policy Purposes." *The Energy Journal* 14, no. 1: 245–255.

Shlyakhter, Alexander I., Clair L. Broido, and Daniel M. Kammen. 1992a. "Quantifying the Credibility of Energy Projections from Trends in Past Data: The U. S. Energy Sector." Department of Physics, Harvard University, processed, October 16.

Shlyakhter, Alexander I., and Daniel M. Kammen. 1992b. "Quantifying the Range of Uncertainty in Future Development from Trends in Physical Constants and Predictions of Global Change." Working Paper 92–106, Global Environmental Policy Project, Harvard University, October 16.

Shlyakhter, Alexander I., and Daniel M. Kammen. 1992c. "Sea-Level Rise or Fall." *Nature* 357 (May 7): 25.

Siegenthaler, U., and H. Oeschger. 1987. "Biospheric $CO_2$ Emissions During the Past 200 Years Reconstructed by Deconvolution of Ice Core Data." *Tellus* 39B: 140–159.

Sohngen, Brent, and Robert Mendelsohn. 1998. "Valuing the Impact of Large-Scale Ecological Change in a Market: The Effect of Climate Change on U. S. Timber." *American Economic Review* 88, no. 4 (September): 686–710.

Stockfisch, J. A. 1982. "Measuring the Social Rate of Return on Private Investment." In Robert C. Lind, ed. *Discounting for Time and Risk in Energy Policy*, Resources for the Future, Washington, D.C., pp. 257–271.

UNDP. 1992. *Human Development Report, 1992*. New York: Oxford University Press.

EIA. 1996. U. S. Department of Energy, Energy Information Administration. *Emissions of Greenhouse Gases in the United States, 1996*. Washington, D.C.

von Neumann, John, and Oskar Morgenstern. 1943. *Theory of Games and Economic Behavior*. Princeton, NJ: Princeton University Press.

Wagonner, Paul E., ed. 1989. *Climate Change and U. S. Water Resources*. New York: Wiley and Sons.

Wang, W.-C., M. P. Dudek, X.-Z. Liang, and J. T. Kiehl. 1991. "Inadequacy of Effective $CO_2$ as a Proxy in Simulating the Greenhouse Effect of Other Radiatively Active Gases." *Nature* 350: 573–577.

Weitzman, Martin. 1974. "Prices v. Quantities." *Review of Economic Studies* XLI: 477–491.

Wuebbles, Donald J., and Jae Edmonds. 1988. *A Primer on Greenhouse Gases.* DOE/NBB-0083. Prepared for the Department of Energy, Washington, D.C., March.

Yang, Zili. 1993. *Essays on the International Aspects of Resource and Environmental Economics.* Ph. D. dissertation, Yale University, New Haven, CT.

Yellen, Janet. 1998. Testimony before Subcommittee on Energy and Power, U. S. Congress House Commerce Committee, March 4.

Yohe, Gary, James Neumann, Paul Marshall, and Holly Ameden. 1996. "The Economic Cost of Sea Level Rise on U. S. Coastal Properties from Climatic Change." *Climatic Change* 32: 387–410.

# Index